Discharge in Long Air Gaps

Modelling and applications

Discharge in Long Air Gaps

Modelling and applications

A Beroual and I Fofana

Ecole Centrale de Lyon, AMPERE Lab, Ecully, France and University of Québec at Chicoutimi, Québec , Canada

IOP Publishing

ISBN 978-0-7503-1236-3 (ebook)
ISBN 978-0-7503-1237-0 (print)
ISBN 978-0-7503-1238-7 (mobi)

DOI 10.1088/978-0-7503-1236-3

Version: 20160601

IOP Expanding Physics
ISSN 2053-2563 (online)
ISSN 2054-7315 (print)

British Library Cataloguing-in-Publication Data: A catalogue record for this book is available from the British Library.

Published by IOP Publishing, wholly owned by The Institute of Physics, London

IOP Publishing, Temple Circus, Temple Way, Bristol, BS1 6HG, UK

US Office: IOP Publishing, Inc., 190 North Independence Mall West, Suite 601, Philadelphia, PA 19106, USA

Contents

Preface

The study of discharge in long air gaps and lightning phenomenon is motivated by (1) the development of High Voltage and Ultra High Voltage power networks where phase-to-ground, and phase-to-phase air gap clearances range from some meters to some tens of meters; (2) the necessary increase in voltage level for long distances; and (3) the protection of people and ground structures.

The discharge phenomena in air gaps have a high practical importance. The knowledge of the breakdown characteristics of many electrode configurations in long air gaps is important to solve a number of practical (engineering) problems arising in outdoor insulation systems. There is therefore a great interest in understanding the behavior of long air gaps breakdown under different kinds of voltage surges and especially lightning and/or switching surges.

The basic physical processes of the discharge in air have been investigated with the help/sponsorship of several experimental programs in the last 40 years. These pioneering experimental investigations have allowed characterisation of the different phases of the breakdown process together with the transition between them. A large number of theoretical investigations dealing with specific modelling of these various phases have been published. These models are concerned with the general descriptions of the basic processes based on some simplifying assumptions, according to the dominant phenomena, which lead to various degrees of complexity in the mathematical formulation.

Theoretical models that simulate discharge processes in long air gaps are useful for two main purposes: (1) they are helpful for a better understanding of the physical processes involved in the discharge; and (2) they accurately simulate discharge processes in long air gaps that are useful for the prediction of breakdown characteristics over a wide range of geometry and atmospheric conditions, therefore reducing the full-scale testing required for the determination air gap clearances of a new transmission line. The ability of some of these models to predict satisfactorily the flashover voltage at any level has already been demonstrated.

Research on the electrical discharge in long air gaps has been primarily directed towards the understanding of the growth of the discharge and the development of mathematical models. Many theoretical studies have been carried out to develop self-consistent predictive models of discharge development. Due to the complexity of the mechanisms involved in air gap discharge, many simplifying assumptions have been made to render mathematical modeling possible. A number of mathematical models describing the physical processes occurring in each discharge phase (first corona, streamer propagation, leader formation and propagation, final jump, return stroke) are available. However, most of these models are limited to single phases and cannot be combined in a general model able to predict the behaviour of a given gap subjected to a given voltage. This is mainly due to the complexity of the phenomena involved in the discharge.

Can one talk about long air gap discharge without any thought of lightning discharge? The prediction of parameters characterising the lightning discharges is of great interest for the protection of people and engineering structures. The availability of mathematical models allowing the prediction of striking distances and the attachment of lightning flashes to grounded structures should allow improvement of engineering systems protection. Contrary to positive lightning for which many theoretical models are available, research to derive models for negative lightning are still in the early stages. This is mainly due to the complexity of mechanisms involved in the negative lightning propagation process. Recall that the study of positive discharges is also important since the strength of the insulation to positive polarity surges is widely used to estimate the required clearances in transmission lines.

This book presents a simplified approach to the different phases of discharge development in long air gaps in both polarities. Furthermore, emphasis is laid on lightning discharge phenomenology. A short review of some of the models proposed by different investigators to simulate air gap discharge is presented. Then self-consistent dynamic models based on equivalent electrical network concepts along with some physical laws, developed by the authors are presented. Based on the similarity between long laboratory sparks and lightning, mathematical models were also developed for positive and negative lightning. The main feature of these models is their ability to predict the characteristics of both positive and negative discharges over a wide range of geometry and atmospheric conditions. Some engineering applications (critical flashover voltage, electromagnetic compatibility, attachment to grounded structures ....) of these models are provided.

Our intention is not to describe in detail the physics of discharge in air which has already been the subject of many books and publications (Rather, Meek and Loeb, Gallimberti, Uman, Raizer – Bazylian and others). The models developed by the authors are based on those pioneering works on the physics of discharge in small and long air gaps. Most material presented in this book has been the subject of papers, published by the authors in various journals and conferences.

Any constructive suggestions will be welcome.

A Beroual
École Centrale de Lyon, University of Lyon, Écully, France

I Fofana
Université du Québec à Chicoutimi, Chicoutimi, Québec, Canada

Acknowledgments

The authors would like to thank various publishers of journals and conference records, in particular *Journal of Physics D: Applied Physics*, *IEEE, IET, ICLP, ISH* and *ICHVE* ... and all editors who allowed us to use previously published materials (figures, diagrams and tables) in this book as noted in the references.

We express our sincere gratitude to Professor Stan Grzybowski from the Missisipi State University, USA for his support and advice.

We are grateful to graduate students who were involved in some aspects of the long air gap and lightning discharges modelling at the High Voltage Laboratory of the Ecole Centrale de Lyon, France. We are indebted to Professor Ph Auriol (now retired from this institution) who supported us and provided all material and help needed.

One of the authors, Professor A Beroual, started his research on long air gap discharges in 1977, in collaboration with Professor S Szpor at the Gdansk University of Technology, Poland. He is grateful to all those from the Institute of High Voltage and Electrical Apparatus who provided their help at that time during experiments.

The authors are also much indebted to the anonymous reviewers for their valuable comments and suggestions which improved the English expressions and technical quality of the manuscript

Finally, we would like to express our gratitude to our family and especially our wives for their invaluable support.

Author biography

Abderrahmane Beroual

Abderrahmane Beroual received the Dipl. Ing. Degree in electrical engineering in 1976 from the Ecole Nationale Polytechnique of Algiers, Algeria; the PhD in high voltage engineering in 1979 from the Technical University of Wroclaw, Poland; and the Doctorat d'Etat ès Sciences Physiques degree in 1987 from the Institut National Polytechnique of Grenoble, France. In 1980, he joined the Department of Electrical Engineering at the Ecole Nationale Polytechnique of Algiers where he initiated the High Voltage research Group. From 1983 to 1987, he was a researcher at the Laboratoire d'Electrostatique et de Matériaux Diélectriques (LEMD) – CNRS of Grenoble and visiting professor at the Joseph Fourier University of Grenoble, France. In 1987, he was appointed full professor at the Ecole Nationale Polytechnique of Algiers where he maintained his research by supervising PhD students. In 1992, he moved to a permanent position at the Ecole Centrale de Lyon – University of Lyon, France where he is currently professor and chair of the Dielectric Materials and High Voltage Group at AMPERE Laboratory – CNRS; he is also Director of the Master Research Program in Electrical Engineering. Professor Beroual was elected IEEE fellow in 2011 'for contribution to processes of pre-breakdown and breakdown in dielectric liquids' He chaired the International Study Group on Streamer Propagation in Liquids of the IEEE – DEIS from 1994 to 1998. He is/was member of many Advisory Committees of International Conferences (ICDL, ISH, APTADM, Gas Discharge, ICHVE, CMD, INSUCON, NIWE, iREED, Technical Committee of the IEEE CEIDP, UF10 Technical Commission – MT30 of IEC) and associate editor for IEEE TDEI. His research work covers a wide range of topics in high voltage engineering and dielectric materials. His research has made a substantial contribution to long air gaps discharge and lightning, outdoor insulation, modelling of discharges and composite materials, pre-breakdown and breakdown phenomena in dielectric liquids and solid/fluid interfaces. He has supervised more than 50 PhD theses. He is author/co-author of more than 400 technical papers including more than 160 refereed journal papers, 5 patents, 1 book and 5 book chapters.

Issouf Fofana

Issouf Fofana received an electro-mechanical engineering degree in 1991 from the University of Abidjan (Côte d'Ivoire), and both a Master's and PhD degree from the École Centrale de Lyon, France in 1993 and 1996, respectively. He was a postdoctoral researcher in Lyon in 1997 and at the Schering Institute of High Voltage Engineering Techniques, University of Hanover, Germany from 1998 to 2000. He was a Fellow of the

Alexander von Humboldt Stiftung from November 1997 to August 1999. He joined the Université du Québec à Chicoutimi (UQAC) as an Associate Researcher in 2000 and he is now Professor there.

Dr Fofana has held the Canada Research Chair, tier 2, of insulating liquids and mixed dielectrics for electrotechnology (ISOLIME), from 2005–15. In 2014, he became Director of the Modeling and Diagnostic of Power Network Equipment (MODELE) laboratory. Since 2015, he is holder of UQAC's chair on the Ageing of Power Network Infrastructure (ViAHT). Dr Fofana's research in the area of HV engineering has the emphases on the on insulation diagnostic/modelling relevant to power equipment. He is registered as a professional engineer in the province of Quebec and, is/was member of many Advisory Committees of International Conferences (IEEE ICDL, ISH, Technical Committee of the IEEE CEIDP) and associate editor for IET Power Transmission and Distribution. From 2013–15, he was appointed to the Executive Committee of the IEEE CEIDP as publication chair (2013) and secretary (2014–15). Since 2016, he is serving the IEEE CEIDP as board member. He is also member of several working groups (CIGRE and ASTM). His lifetime publication record is more than 200 scientific publications and 3 patents.

IOP Publishing

Discharge in Long Air Gaps
Modelling and applications
A Beroual and I Fofana

Chapter 1

Introduction

1.1 Research into and interest in the long air gap discharge

The study of electrical discharges in long air gaps is of interest both in terms of physical applications (atmospheric physics, plasma physics, etc) and from the industrial point of view (air gap clearances in high voltage (HV) engineering). This interest is closely linked to the development of overhead transmission lines over long distances. The transport of large amounts of electrical power over long distances is achieved with HV transmission lines, to reduce current and therefore line losses (Joule heating), and line voltage drops, and to improve network stability.

Historically, the study of HV transmission technologies began in the United Kingdom, Russia, Japan, the USA, Italy, Canada, and Brazil in the 1960s and 1970s [1]. With a growing world population and faster developing industries, larger amounts of electric power need to be transferred longer distances, which requires higher transmission voltages. Joule losses are proportional to the resistance of the conductor and the square of the current. Thus, to reduce losses, the voltages delivered by generators are increased with step-up transformers, and electric energy is transported across the countryside by HV lines. With increased voltages, large air gap distances between phases (of a few to more than ten metres) are required. Many ultra/extra-HV transmission projects have been commissioned or are under construction in many countries, e.g. Canada (735 kV), Venezuela (800 kV), China (1100 kV), Japan (1100 kV) and India (1200 kV).

The dielectric strength of insulation systems essentially depends on the inter-electrode distance, the electrode shape, the shape and polarity of the voltage, and the atmospheric conditions. Because of the relation between the dielectric strength of air and the mechanisms involved in long discharges, this phenomenon has long been the subject of intensive study in many laboratories worldwide [1].

The first work on discharges in long gaps was carried out in the 1920s under an ac voltage. The recordings of over-voltages in electrical networks showed that a lightning current wave presents a unidirectional pulse, the shape of which is bi-exponential with a very fast front and a slower decreasing tail. In the 1930s, the shape of the lightning

doi:10.1088/978-0-7503-1236-3ch1

impulse was standardized and the first multistage impulse generator designed by E Marx (back in 1923) became a convenient source for testing materials/equipment. These developments helped in the establishment of HV laboratories worldwide and intensive experimental studies, including those reported on long discharges, were undertaken.

The photography of lightning discharges, undertaken in 1934 in South Africa by B Schonland and H Collens [2] using a Boys camera, enabled the recording of images depicting the characteristics of lightning discharges and the gathering of important data [2]. They confirmed that a lightning discharge propagates step-wise with two main phases: the leader and the return stroke phases. The existence of these two phases was confirmed by B Allibone and T Schonland under laboratory conditions for a gap of 2 m [3].

Studies of long discharges for gaps of over 2 m were pursued in many research centres, mainly in the UK and the former USSR [4–8]. This research focused on the physical phenomena leading to the initiation and propagation of discharges at the origin of the insulation system breakdown. Different parameters characterizing the discharge, such as the leakage current and velocity, were investigated. These studies have also shown that the mechanism of discharge development in large gaps is different from Townsend's theory [9] wherein, under normal atmospheric conditions, discharge is verified only for short electrode gaps (of about 1 cm). In Townsend's theory, electron emission by photo-ionization at the cathode plays an important role in the inception of a self-sustained discharge. In contrast, in long air gaps the cathode does not influence the further development of the discharge; in a non-uniform electric field the long discharge develops at electric field strengths much lower than those deduced from Townsend's theory.

To overcome the fundamental shortcomings of Townsend's theory in explaining the phenomena observed in long air gaps, Meek and Loeb [8, 10–12] proposed the first theoretical work on the mechanism of the channel created by the long discharge, the *streamer–leader theory*. In this theory, photo-ionization resulting from excitation phenomena and recombination in the gap as well as space charge build-up are the fundamental processes in the development of discharges. The authors showed that a plasma channel leader is created after the streamer phenomena. A streamer is a narrow filament of ionized gas, resulting from the connection and the penetration of several secondary avalanches into the region of the primary electron avalanche, reaching the electrode surface. This mechanism causes the creation of an electrode charge at the head of the streamer and the elongation of the filament. The photo-ionization and amplification phenomena of the field in front of the avalanche play an important role in the generation of new avalanches [8, 10–13]. The number of streamer filaments, in particular in the vicinity of the anode, are large and occupy a large area.

Using a Boys camera, V Komelkov estimated the velocities of the streamers and the leader's tip, and the leader diameter, moving from a rod electrode to a grounded plane electrode [6]. He reported that the streamer velocity (about 1 m μs^{-1}) is very high, compared to that of the leader tip (about 1 cm μs^{-1}), while the diameter of the leader is about 1 mm. In 1978, S Szpor, H Boryn, and A Beroual [14] investigated the influence of a series resistance (a resistance inserted between the HV generator and a sphere–sphere spark gap) on the diameter of discharge under a standard lightning voltage. They observed that the diameter of the discharge channel as well as the accompanying emitted light decreased when the value of the resistance increased. Other researchers estimated these parameters more accurately using new techniques of recording and measuring very fast phenomena [15, 16].

Considering a 2 m spark gap consisting of a rod ending in a small sphere and a grounded plane, R Waters and R Jones [17, 26] observed the existence of a time lag between the inception of the first corona and the leader initiation. They concluded that the electric charge injected into the inter-electrode gap by the streamers slows down the development of the discharge.

In the 1960s, many researchers investigated the dielectric strength of long air gaps under positive impulse voltages, considering different front rise times, ranging from a few microseconds to a few hundred microseconds [8–13, 17–26]. They showed that by increasing the rise time, the dielectric strength of the electrode gap first decreases and then increases, indicating the existence of a critical front rise time of the voltage at which the dielectric strength of the insulation system is the lowest. This result is of great interest in practice, in particular for the design of large air gap insulation systems.

The first pictures of discharges obtained using an image converter (or optoelectronic converter) were reported by S Stelkolnikov and A V Shkilev [27, 28]. In 1962 they published very interesting photographs that helped to significantly progress the understanding of the various stages of discharge propagation and accurately track the emitted light.

The understanding of discharge phenomena has been greatly enriched thanks to the use of new optical techniques and very fast measurement apparatus, including fast optical detecting systems using Shadowgraph and Schlieren photography coupled with an image converter camera, fast digitizing oscilloscopes for voltage and current measurements, electric field distribution measuring systems, optical spectroscopy of emitted light, etc. In the 1960s and 1970s, J Meek and M Collins [29], R Waters [30], and R Waters and E Slim [31] developed methods which allow the measurement of the intensity of the electric field at the surface of the HV electrode.

In the 1970s an international group known as the 'Les Renardières Group' emerged, composed of researchers from several countries [32–36]; the name of this group came from the fact that the tests were conducted in the French laboratory Les Renardières. The objectives of this group were to conduct measurement campaigns to determine the important parameters characterizing long air discharges with gaps ranging from a few centimetres to more than ten metres (~17 m), and to improve our basic understanding of the processes involved. Different electrode geometries (rod–plane, sphere–sphere, and cylinder–plane electrode systems) subjected to positive and negative switching impulse voltages were investigated. Similar studies were also conducted in other laboratories worldwide. This allowed the determination of the fundamental parameters of long air discharges (the current and electrical charge, the streamer and leader velocities, the size of the discharge channel, the voltage drop at the discharge channel, the electric field, etc) and the description of the physical mechanisms involved in the different phases of discharge development [32–40].

From this work, the phenomena involved in long air discharges were better understood and the resulting data were used to develop different models to describe the various phases of development of the discharge. Many theoretical works dealing with the specific modelling of the different phases [36, 38, 41–45] as well as models describing the whole discharge [46–51] have been published. These models are based on descriptions of the basic physical processes and state some simplifying assumptions, according to the dominant phenomena, which lead to various degrees of complexity in the mathematical formulations. Also, thanks to studies conducted by Les Renardières, the similarity between the long air discharge and lightning was established [52] allowing the proposal of models [53, 54] that helped to improve the design of HV equipment/hardware along with lightning protection concepts.

A large number of studies have been reported in the literature. The list of contributions that should be referred to is very long. The aim of this brief historical review on the long discharge is to provide an overview of the significant progress in the understanding of the long discharge phenomenon. This has enabled researchers to acquire rich experimental data on the long discharge. Theoretical studies based on these data have helped to explain many peculiarities of the process of discharge formation. However, despite the enormous progress in the understanding the phenomenon of long discharge, many issues still require deeper study.

This book is directed toward the mathematical modelling of long air gap discharges. Before entering the heart of this book, we recall first the main mechanisms related to discharge phenomena. Our intention is not to compete with existing books on the physics of discharge. The books published by Meek and Craggs in 1953 [41], and more recently by Bazelyan and Raizer (1997), constitute, in our opinion, the fundamental publications available on long discharges in which one can find the basic elements that describe the physicochemical processes involved in the different phases of discharge. They give the reader a detailed review of the experimental and theoretical work on long sparks. Finally, note that we use the experimental results published by different authors, and in particular those of Les Renardières Group, to validate the models described in this book.

1.2 Scope and objectives

Most hardware used in HV systems requires a sound geometric design for air gap clearance. The required insulation distances are carefully determined to withstand surge over-voltages. Some of the common over-voltages occurring on power transmission lines include the following [55]:

- Temporary over-voltages that are caused by faults, load rejection, line energizing, resonance conditions, ferroresonance, or by some combinations of these factors.
- Switching (slow-front) over-voltages that are caused by switching operations, fault initiation, or remote lightning strikes.
- Lightning (fast-front) over-voltages that are caused primarily by lightning strikes but could also be caused by some switching operations or fault initiation.
- Very fast-front over-voltages that are the result of switching operations or faults and are usually associated with power switch operations, gas insulated switchgears, and cable connected motors.

Air gap clearance design specifications, based on empirical and probabilistic methods and reported in the standards [55–57], are developed to minimize the risk of flashover/breakdown at a reasonable cost. Understanding the fundamental theoretical causes of air gap breakdown is therefore important for the proper design of practical HV transmission lines and hardware. Knowledge of the physical mechanisms that occur in the breakdown of long air gaps and the parameters that characterize this discharge are thus necessary to solve a variety of engineering problems, in particular for insulation designs that use air as the insulating medium [58, 59].

Another important objective that motivates research on large air gap discharge phenomena is the improvement of our basic knowledge of lightning discharge. It should be noted that a substantial similarity exists between the lightning phenomenon and the discharge in long air gaps [48, 52, 60–62].

This book is mainly focused on the modelling of electrical discharge in long air gaps. Much attention is given here to the mathematical modelling of the discharge processes. This book is also intended to provide a fresh perspective by contributing an innovative approach to this research domain.

1.3 Intended audience

Reliability and power quality specialists for HV networks often question why hardware insulated with an air gap fails. It does so because the design and dimensioning of adequate air gap clearance coupled with good performance under various atmospheric conditions is still a challenge. This problem is partly due to the complexity of the discharge phenomena. Based on an appropriate physical theory of the electric discharge process, dynamic modelling may contribute to a better understanding of the phenomena involved in the whole range of discharge mechanisms. Although considerable effort has gone into the characterization of discharge phenomena in the last few decades, relatively few studies have been carried out to develop comprehensive mathematical models in this area.

This book is primarily directed towards the mathematical modelling of long air gap discharges. Predictive dynamic models that simulate large air gap discharges under positive and negative impulse voltages are presented. Based on the similarities between long air gap discharges and lightning flashes, equivalent models are also derived to predict lightning parameters. Macroscopic air gap discharge parameters are calculated by computer programs that solve electrical, empirical, and physical equations. Good agreement was obtained between the computed and experimental results for various test configurations.

Universities with programs on HV engineering will find this volume to be a working example of how to introduce the basics of electric discharge phenomena. For engineering programs with a power system emphasis, this book may serve well as a graduate-level textbook for special topics in HV electrical engineering, insulation coordination, or other courses on power engineering related to power transmission and distribution. The study of electrical discharge in large air gaps is also a compelling and accessible introduction to lightning discharge and, therefore, provides a better understanding of the attachment of lightning flashes to ground structures as well as the way in which a radiated electromagnetic field may affect power equipment or hardware.

References

[1] Candas M and Meric O S 2015 *J. Power Energy Eng.* **3** 453

[2] Schonland B F J and Collens H 1934 *Proc. R. Soc. Lond.* **143** 654

[3] Allibone T E and Schonland B F J 1934 *Nature* **134** 735

[4] Allibone T E and Meek J M 1938 *Proc. R. Soc. Lond.* A **166** 97

[5] Stekolnikov I S, Beljakov A P and Mjakishev I N 1937 *Elektrichestvo* **8** 51

[6] Komelkov V S 1947 *Bull. Acad. Sci. USSR* **8** 955

[7] Komelkov V S 1950 *Bull. Acad. Sci. USSR* **6** 851

[8] Loeb L B and Meek J M 1941 *The Mechanism of Electric Spark* (Oxford: Clarendon)

[9] Townsend J S 1941 *Electricity in Gases* (Oxford: Oxford University Press)

[10] Loeb L B 1960 *Basic Processes of Gaseous Electronics* (Berkeley, CA: University of California Press)

[11] Loeb L B 1965 *Electrical Coronas: Their Basic Physical Mechanisms* (Berkeley, CA: University of California Press)

[12] Meek J M 1969 *Proc. IEEE* **116** 1

[13] Raether H 1964 *Electron Avalanches and Breakdown in Gases* (London: Butterworths)

[14] Szpor S, Boryn H and Beroual A 1978 *Arch. Elektrotech.* **3** 483

[15] Ross J N 1977 *Electra* **53** 71

[16] Ruhling F 1977 *Electra* **53** 67

[17] Waters R T and Jones R E 1959 Time-resolved photography of arrested discharges in long gaps *Proc. Fourth Int. Conf. on the Phenomena of Ionization in Gases* (*Uppsala, Sweden, 17–21 August*) p 408

[18] Waters R T 1962 Streak photography and other studies of the long spark in air *Gas Discharges and Electricity Supply Industry* (London: Butterworths) p 38

[19] Waters R T and Jones R E 1964 *Phil. Trans. R. Soc.* A **256** 185

[20] Waters R T and Jones R E 1964 *Phil. Trans. R. Soc.* A **256** 2123

[21] Bazelyan E M, Brago E N and Stekolnikov I S 1961 *Sov. Phys. Dokl.* **5** 794

[22] Stekolnikov I S and Shkilev A V 1964 *Sov. Phys. Dokl.* **8** 825

[23] Wada S 1970 *Electr. Eng. Japan* **90** 58

[24] Carrara G, Dellera L and Satorio G 1970 *IEEE Trans. Power Appar. Syst.* **89** 453

[25] Jones B and Whittington H W 1972 *IEEE Conf. Pub.* **90** 39

[26] Jones B and Whittington H W 1972 Influence of the shape of the testing waveform on breakdown of long air gaps *Int. Symp. on High Voltage Engineering* (*Munich, Germany, 9–14 March*) p 292

[27] Stekolnikov I S 1962 *Sov. Phys. Dokl.* **6** 1085

[28] Stekolnikov I S and Shkilev A V 1961 *Dokl. Akad. Nauk USSR* **136** 803

[29] Meek J E and Collins N M C 1965 *Electr. Lett.* **4** 110

[30] Waters R T 1972 *J. Phys. E: Sci. Instrum.* **5** 475

[31] Waters R T and Slim E O 1979 Field filters: new static probes for field measurements at electrodes *Third Int. Symp. on High Voltage Engineering* (*Milan, Italy, 28–31 August*) 44-07

[32] Les Renardières Group 1972 *Electra* **23** 53

[33] Les Renardières Group 1974 *Electra* **35** 49
[34] Les Renardières Group 1977 *Electra* **53** 31
[35] Les Renardières Group 1981 *Electra* **74** 67
[36] Les Renardières Group 1986 *Proc. IEEE* A **133** 395
[37] Aleksandrov G N 1966 *Sov. Phys.—Tech. Phys.* **10** 948
[38] Gallimberti I 1979 *J. Physique Coll.* **40** 193
[39] Allibone T I 1977 The long spark *Lightning* ed R H Golde (London: Academic) p 231
[40] Marode E 1975 *J. Appl. Phys.* **46** 2005
[41] Meek J M and Craggs J D 1953 *Electrical Breakdown of Gases* (Oxford: Clarendon Press)
[42] Dawson G A and Winn W P 1965 *Z. Phys.* **183** 59
[43] Gallimberti I 1972 Mathematical model for streamer model in non-uniform gaps *Padova University Report* Upee 72/03
[44] Hutzler B and Hutzler D 1992 *Bull. DER d'EDF* **4** 11
[45] Bondiou A and Gallimberti I 1994 *J. Phys. D: Appl. Phys.* **27** 1252
[46] Fofana I and Beroual A 1995 *J. Phys. D: Appl. Phys.* **28** 305
[47] Fofana I and Beroual A 1996 *J. Phys. D: Appl. Phys.* **29** 691
[48] Fofana I and Beroual A 1996 *IEEE Trans. Dielectr. Electr. Insul.* **3** 273
[49] Fofana I and Beroual A 1997 *J. Phys. D: Appl. Phys.* **30** 1653
[50] Beroual A 2004 *REE Rev. Electr. Electron.* **8** 36
[51] Rakotonandrasana J H, Beroual A and Fofana I 2008 *J. Phys. D: Appl. Phys.* **41** 105210
[52] Hutzler B 1988 Comparaison foudre—étincelle de laboratoire *Texte de Communications SE, Groupe: Décharges Électriques* p 21
[53] Beroual A, Rakotonandrasana J H and Fofana I 2010 *IEEE Trans. Dielectr. Electr. Insul.* **17** 1551
[54] Beroual A, Rakotonandrasana J H and Fofana I 2010 *IEEE Trans. Dielectr. Electr. Insul.* **17** 1562
[55] IEEE 1999 *IEEE Guide for the Application of Insulation Coordination standard* 1313.2-1999
[56] IEC 1996 *Insulation Coordination—Application Guide* part 2 3rd edn (Geneva: IEC) 600071-2-1996
[57] EPRI 1982 *Transmission Line Reference Book: 345 kV and Above* 2nd edn (Palo Alto, CA: EPRI)
[58] Hill K H 1999 Estimation of positive polarity switching surge flashover voltage of compact transmission lines *MSc Thesis* University of Manitoba, Winnipeg
[59] IEEE 1995 *IEEE Standard Techniques for High Voltage Testing* Standard 4-1995
[60] Reess T 1997 Contribution à l'étude expérimentale et théorique des décharge électriques de polarité négative dans l'air à pression atmosphérique *PhD Thesis* Université de Pau, France
[61] Ortéga P 1992 Comportement diélectrique des grands intervalles d'air soumis à des ondes de tension de polarité positive ou négative *PhD Thesis* Université de Pau, France
[62] Hutzler B 1988 *Bull. Dir. Etud. Rech.* **3** 31

IOP Publishing

Discharge in Long Air Gaps
Modelling and applications
A Beroual and I Fofana

Chapter 2

The background of air gap discharge theory

2.1 Introduction

The importance of insulating gases has been acknowledged from the early days of the study of electricity. Air is abundant and has been used since the beginning of HV engineering in electrostatics, transmission lines, etc. When subjected to a voltage gradient, air may conduct electricity, depending on the magnitude of the applied voltage. At a relatively low voltage, the flow of current is negligible. The flow of current will increase as voltage increases. When the applied voltage gradient reaches a certain critical value (relative to air), an electrical discharge (or air breakdown) occurs: the gas switches from a non-conducting to a conducting medium and consequently short-circuits the electrodes. The comprehension of the basic processes from the origin of the initiation of these discharges up to breakdown is fundamental. The study of these processes has been the subject of a large number of books and publications (e.g. [1–8]). Basically, the fundamental processes that are primarily responsible for air gap breakdown are ionization by collision, photo-ionization, thermal ionization, and secondary ionization processes. Other processes, such as electron attachment, recombination, and diffusion, also affect the discharge mechanisms. The generation of electrons and ions from neutral atoms or molecules and their migration to the anode and cathode, respectively, will contribute to increasing the current flow with any increase in the electric field (applied voltage). The various physical parameters of the air gap, i.e. pressure, temperature, electric field stress, the material and configuration of the electrodes (geometry, gap distance), and the presence of initial conducting particles, are known to influence the ionization processes.

Two types of electrical discharge that depend on the voltage magnitude and, hence, on the electric field can be distinguished: (i) non-sustaining and (ii) self-sustaining. Breakdown of the air gap, also known as spark breakdown, is the transition from a non-sustaining discharge to a self-sustaining discharge. Together with the applied voltage, the gap distance pays an important role in the discharge process. For air gaps below one centimetre, the accepted theory has been proposed by Townsend, and is

doi:10.1088/978-0-7503-1236-3ch2

known as 'Townsend theory' [9]. However, for air gaps of a few centimetres or more, the most used theory is that proposed by Loeb and Meek [1], known as 'streamer–leader theory'.

This chapter is intended to set the scene for more detailed discussions about the physics of discharge in air. Some fundamental processes, including kinetic theory and the breakdown phase of a direct current small air gap discharge, are recalled. Because this textbook is mainly focused on the mathematical modelling of large gaps and some relevant applications, only the most significant processes are presented.

2.2 Ionization phenomena

2.2.1 Ionization by collision

Ionization by electron impact is probably the most important process governing the breakdown of gases. This process consists of liberating an electron from a gas molecule/atom M simultaneously producing a positive ion.

$$e^- + M \rightarrow e^- + M^+ + e^-, \tag{2.1}$$

where M is the atom, M^+ is the positive ion, and e^- is the electron.

Under the electric field E, any charges between the electrodes will experience a Lorentz force, conveniently combining the electric and magnetic effects [10]:

$$F = q(E + v \wedge B), \tag{2.2}$$

in which q is the quantity of charge in coulombs; E is the electric field in volts per metre; v is the velocity in metres per second; $\wedge$ implies the vector product; and B is the flux density in tesla.

If the particle (electron or ion) of mass m is moved at a drift velocity v, it gains the energy W_k:

$$W_k = \frac{1}{2}mv^2. \tag{2.3}$$

The energy gained by moving on a given distance, x_1, in the direction of the electric field E, can be expressed as

$$eV = \int_0^{x_1} qE\mathrm{d}x. \tag{2.4}$$

The effectiveness of atom/molecule ionization by collision depends on the kinetic energy (W_k). It is possible only when the kinetic energy is equal to or higher than the ionization energy $W_i(W_k \geqslant W_i)$. Most of the ionization process arises from high energy impacts between electrons and atoms/molecules.

The collision of a fast electron with a gas atom/molecule M may be either elastic or inelastic, leading to very different results [7]:

(i) *Elastic collisions.* When elastic collisions occur, no change takes place in the internal energy of the particles, and the kinetic energy is only redistributed between particles so that the total kinetic energy remains unchanged, e.g.,

$$e^-_{\text{fast}} + M_{\text{slow}} \rightarrow e^-_{\text{less fast}} + M_{\text{less slow}}. \tag{2.5}$$

Lighter particles (M_L) cannot lose much energy elastically to heavier particles (M_H)—at best a fraction $2M_L/M_H$; nevertheless, substantial changes in momentum occur (as in a tennis game) [7]. Since electrons are very light, only a fraction of their kinetic energy is transferred to the much heavier ions or gas molecules with which they collide. Very little energy is lost by the electrons; therefore, electrons gain very high energy and travel at a much higher speeds than the ions. Consequently, electrons play a leading role in all electrical discharges through elastic collisions.

(ii) *Inelastic collisions*. In inelastic collisions momentum is redistributed between particles, but a fraction of the initial kinetic energy is transferred to internal energy in one or more of the particles (i.e. excited states or ions are formed), e.g.,

$$e^{-}_{fast} + M_{slow} \rightarrow e^{-}_{slower} + M^{*} \text{ (excited molecule)} \tag{2.6}$$

or

$$\rightarrow e^{-}_{slower} + M^{+} + e^{-}. \tag{2.7}$$

These collisions do not occur in practice.

If the energy (eV) gained during electron travel exceeds the ionization potential, V_i, (the energy required to dislodge an electron from its atomic shell), then ionization takes place according to equation (2.1) or (2.7).

2.2.2 Photo-ionization

Free electrons may be released by some external means (natural or artificial ionizing radiation, i.e. photons, ultraviolet (UV) light of a suitable frequency). For an electron to escape from a particle (atom/molecule), it should be given enough energy to overcome the surface potential barrier. Since ionization energies tend to be two or three times the value of the work functions, photo-ionization is always associated with deep UV radiation, except when it involves interactions between photons and neutral particles already excited to (temporary) higher states [7]. In this case, photo-ionization consists of ionizing atoms or molecules by radiation. This happens when a photon's energy absorbed by the atom or molecule is higher than its ionization energy:

$$M + h\nu \rightarrow M^{+} + e^{-} \tag{2.8}$$

$$M^{*} + h\nu \rightarrow M^{+} + e^{-}. \tag{2.9}$$

Another source of electron emission is surface photo-ionization at plasma boundaries. Photons are effective removers of electrons provided that their energy $h\upsilon$ exceeds the work function (ϕ) of the metal (the so-called surface ionization); the excess energy is present in the kinetic energy of the emitted electrons. Planck's constant is represented by $h = 6.626 \times 10^{-34}$Js, while υ denotes the light frequency.

2.2.3 Thermal ionization

Thermal ionization may take place through kinetic energy from the random motion of particles. The average kinetic energy of particles is related to their kinetic temperature:

$$W_k = W_{th} = \frac{1}{2}mv^2 = \frac{3}{2}kT, \tag{2.10}$$

where k is Boltzmann's constant ($k = 1.38 \times 10^{-23}$ J/K) and T is the temperature of the relevant kind of particle.

The kinetic energy is $W_k = 0.038$ eV at $T = 293$ K. This value is two times lower than the ionization energy of most gases. At low temperatures the likelihood of such ionization is, therefore, very low. Since 1 eV is equivalent to a kinetic temperature of 11 000 K [7], any thermal ionization below a few thousand kelvins is not expected. Thus, this kind of ionization plays a fundamental role in the thermalization phase of the discharge, i.e. during the transition to arc (spark).

2.2.4 Ionization by interaction of a meta-stable particle with an atom/molecule

Meta-stability refers to a particular excited state of an atom, nucleus, or other particle that has a longer lifetime than ordinary excited states. The parameters of such 'excited' systems may eventually reach and hold stationary values (a meta-stable state) but then, after a long time, (spontaneously or under a slight external action) parameters will start changing again. A meta-stable particle (a *long-lived* excited particle but not truly indefinite stability) generally has a shorter lifetime than the lowest, often stable, energy state, called the ground state.

The internal energy of meta-stable particles is an important source of energy for electron emission from surfaces at plasma boundaries [7]. When the excitation energy of a meta-stable particle H, is lower than the ionization potential of an atom/molecule M, the excitation energy may be transferred; this process leaves the meta-stable particle in its ground state and the molecule in a highly excited state.

In contrast, when the excitation energy of H is greater than the ionization potential of M, Penning ionization becomes an important source of ionization among species [7]:

$$M + H^* \rightarrow M^+ + H + e^-. \tag{2.11}$$

2.3 Cross section and mean free path—elastic collisions

The collision event is a statistical process that is summarized in terms of how far an electron travels, on average, before it collides with a neutral particle.

Consider an electron moving through a cuboid of gas. This cuboid is assumed to be composed of a given number of stationary gas atoms (see figure 2.1). The atoms are viewed as hard spheres while the electron is assumed to behave as a point mass.

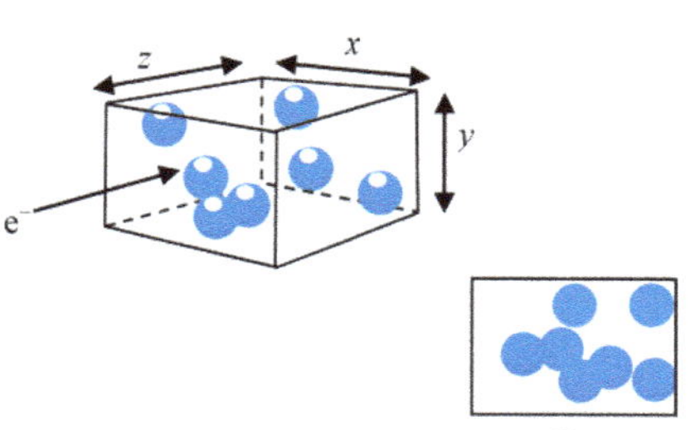

Figure 2.1. Calculation of mean free path, reproduced from [7] with permission.

Effects arising from its charge are not considered. How far does an electron travel on average before a collision? Only simple elastic encounters will be dealt with here.

Let us call λ the *mean free path*, i.e. the average distance travelled by the electron between two collisions in the field direction.

This parameter (λ) is inversely proportional to the number of collisions (n) encountered by the electron during its travel. Any collision process between two particles can be characterized in terms of a 'cross-sectional collision', i.e. in terms of effective cross-sectional area. Each atom in the cuboid presents a cross section, $\sigma = \pi r^2$, obscuring the electron path (r being the radius of the atom).

If N represents the total number of atoms per unit volume in the cuboid, the product $N\sigma$ is the sum of the effective cross sections of all the N target particles in a volume and is called the total effective cross section, absorption coefficient, or macroscopic cross section.

The free path is a random quantity and its mean value depends on the concentration of particles or the gas density. When the cuboid extends as far as one mean free path (λ), virtually the entire face xy will be obscured so that

$$\lambda = \frac{1}{N\sigma}. \tag{2.12}$$

The pressure, p, in a gas is a measure of the density of the thermal energy associated with the number of gas atoms per unit volume, N. Thus,

$$p = NkT. \tag{2.13}$$

Since N is proportional to pressure and inversely proportional to absolute temperature, it yields

$$\lambda = \frac{1}{Ap}, \tag{2.14}$$

where $A = \sigma/kT$ depends on the gas itself. Table 2.1 provides some values of λ for particular gases under normal temperature and pressure conditions.

2.4 Mobility, diffusion, and recombination

In addition to ionization processes that increase the number of free charge carriers in the gas, two other processes, diffusion and recombination, take place, contributing to the reduction in charge carrier density. The evolution of both processes depends on the mobility of charge carriers.

Table 2.1. Free mean path λ of different gases under normal temperature and pressure conditions [11].

Gas	Particle diameter (m)	Mean free path λ (m)
H_2	2.8×10^{-10}	6.5×10^{-7}
N_2	3.8×10^{-10}	3.5×10^{-7}
O_2	3.6×10^{-10}	3.9×10^{-7}
SF_6	4.0×10^{-10}	3.2×10^{-7}

2.4.1 Mobility of electrons and ions

In the absence of an electric field, electrons and ions in the gas are subjected only to random thermal motion and collide with neutral particles. Upon application of the electric field, the balance of forces involves the effect of the electric field (qE, q being the charge of the particle) on the charged particles. The motion of these particles is, therefore, a combination of random thermal activity and the steady drift (v), superimposed by electric fields. In the initial breakdown phases of a gas discharge (preceding the arc phase), thermal activity is negligible; one is thus interested in a few charged particles, accelerated by the electric field (E) interacting with a background gas. The steady drift (v) is then related to the electric field (E) by the mobility (μ)

$$v = \mu E. \tag{2.15}$$

Strictly speaking, it is a function of the particle energy, but simple models consider it as a constant [7].

The mobility of ions (μ_i) can be expressed as:

$$\mu_i = A_p \frac{q\lambda_i}{m_i v_{th}}. \tag{2.16}$$

A_p is a proportionality coefficient ranging from 0.5 to 1; λ_i is the free mean path of the ion; m_i is the mass of the ion; and v_{th} is the average thermal velocity.

Knowing the free mean path of ions λ_i, given by equation (2.12), their mobility can be deduced as

$$\mu_i = A_p \frac{q}{m_i v_{th} A_p}. \tag{2.17}$$

The mobility of ions (μ_i) is directly proportional to the free mean path. Therefore μ_i depends on all the parameters affecting the free mean path. The higher the pressure, the lower the mobility will be (equation 2.12).

The mobility of electrons (μ_e) can be expressed as a function of the ion mobility (μ_i) as [1–4, 11–13],

$$\mu_e = \sqrt{\frac{m_i}{m_e}}\,\mu_i, \tag{2.18}$$

where m_e is the mass of the electron. Therefore, the mobility of electrons is about one thousand times higher than that of ions. Table 2.2 provides some average values for electron and ion mobility.

Table 2.2. Electron and ion mobility.

Particle	Mobility ($cm^2\ Vs^{-1}$)
Electron	2000
Small positive ion	1.37
Small negative ion	1.91
Large ion	10^{-5}

The discharges in gases represent a low path for current conduction. The current density (J) depends on the mobility of the charged particles,

$$J = v_+ n^+ q_+ + v_- n_- q_- + v_e n_e q_- = \left(\mu_+ n^+ q_+ + \mu_- n^- q_- + \mu_e n_e q_-\right), \tag{2.19}$$

where

- $v_{+,-}$ = velocity of positive or negative ions;
- v_e = velocity of electrons;
- n_e = density of electrons;
- $n^{+,-}$ = density of a positive or negative particle;
- $q_{+,-}$ = charge of a positive or negative particle = $\pm 1.6 \times 10^{-19}$ C;
- $\mu_{+,-}$ = mobility of a positive or negative particle;
- μ_e = mobility of electrons.

Electrons are the dominant carriers, unless they are strongly impeded by attachment, recombination, or magnetic fields. The current density (J) is then given locally by

$$J = n_e q_- v_e = n_e q_- \mu_e E. \tag{2.20}$$

2.4.2 Diffusion

Diffusion, i.e. the scattering of ions and electrons, consists of the migration of particles due to density gradients. Particle dispersion depends on the diffusion coefficient (D), which, in turn, is proportional to the free mean path of the particle and to its average thermal velocity

$$D = \frac{\lambda_i v_{\mathrm{th},i}}{3}. \tag{2.21}$$

Kinetic theory allows for linking diffusion to mobility through the Einstein relation

$$D/\mu = kT/q. \tag{2.22}$$

Due to their large mean free path and velocity, the diffusion coefficient of electrons far exceeds that of ions. This may give rise to space charges (i.e. a zone with an excess of unipolar charge carriers). The generation of such charge carriers plays an important role in the further development of discharges by affecting the electric field distribution in the electrode gap.

Basically, diffusion is independent of electrostatic forces. This is mainly due to the fact that the average distances between ions at common concentration rates are too large. Diffusion depends upon another phenomenon known as partial pressure. Equilibrium in the partial pressure of ions means that, after a certain time, a homogeneous distribution of ions is achieved within the gas. Particle dispersion leading to balanced concentration in the gas can be estimated as a function of the diffusion coefficient (D).

2.4.3 Recombination

The reduction of ion concentration within the gas is due not only to the diffusion phenomenon but largely to recombination phenomena. Basically, recombination can be seen as the reverse process of photo-ionization. The energy ($h\upsilon$) released during the electron–ion or ion–ion pair recombination process is the sum of the electron's kinetic energy ($\frac{1}{2}m_e v_e^2$) and ionization energy (W_i) of the gas particle under consideration.

$$h\upsilon = W_i + \frac{1}{2}mv_e^2. \tag{2.23}$$

Simple electron–ion recombination is not a straightforward process. Indeed, due to the high kinetic energy gained by electrons under the field E, the likelihood of positive ion–electron recombination is very low. The photo-ionization route involves very little momentum, so the reverse process requires virtually stationary electrons and ions. During the discharge process, the electron impact ionization route produces three particles. Hence, the reverse process would need the reaction of three bodies to conserve energy and momentum [7]

$$e^- + e^- + A^+ \rightarrow A + e^-_{faster}. \tag{2.24}$$

The recombination is eased through attachment followed by an ion–ion recombination in a second step:

$$e^- + AB \rightarrow AB^- \tag{2.25}$$

$$AB^- + C^+ \rightarrow AB + C. \tag{2.26}$$

The energy released during this process can be expressed as

$$h\upsilon = W_i - W. \tag{2.27}$$

W is the energy required to remove an electron from a negative ion.

The concentration of ions is time-dependent according to the following relationship:

$$\frac{dn^+}{dt} = \frac{dn^-}{dt} = k_r n^+ n^-, \tag{2.28}$$

where n^+ and n^- are the concentrations of positive and negative ions, respectively and k_r is the recombination coefficient of ions.

Assuming $n^+ = n^- = n_0$ (which is generally the case) and $n = n_0$ at $t = 0$, it would yield

$$n(t) = \frac{n_0}{1 + k_r n_0 t}. \tag{2.29}$$

2.4.4 Effect of humidity

From experimental observation, it is known that humidity retards streamer development in air [14]. Indeed, water vapour behaves as an electronegative gas, limiting the total charge injected into the gap by the streamer channels. In such a case, the voltage required to bridge the gap is increased. This phenomenon is observed in highly divergent fields. Under a uniform electric field, the influence of moisture is practically negligible.

2.5 Discharge in small air gaps: Townsend's discharge theory

Consider an air gap column at a given pressure subjected to an electric field (E) across two planar parallel electrodes, as shown in figure 2.2. Any electron located near the cathode will be accelerated (between collisions with other gas molecules) during its travel towards the anode. Positive ions (M^+) formed due to ionization by collision travel towards the cathode, while electrons travel to the anode. The number of these charge carriers increases under the influence of the electric field. The displacement of charge carriers induces a discharge current flowing through the electrodes. This induced discharge current (figure 2.3) is influenced by electrons released from the cathode. Depending on several parameters, such as the degree of ionization, the intensity of the electric field (E), and the gas density (pressure and temperature), various phenomena may be observed, as depicted in figure 2.3.

The evolution of the discharge current as a function of the applied voltage is represented in figure 2.3. This voltage–current (U–I) characteristic shows different regions, each corresponding to a specific phase of the discharge. This phenomenon was first observed over a century ago by Townsend (1900) [11]. These regions (or regimes) are briefly described in the following.

2.5.1 Non-self-sustaining discharge without electronic multiplication

This phase is characterized by a very low current (10^{-14} to 10^{-10} A) without observation of glow phenomena. The applied voltage (a few tens of volts) and, consequently, the applied field are too low to enhance ionization processes by inelastic collisions. Conduction phenomena may take place only by ionization provided by neutral gas particles ionized by external means (for example, by UV light falling on the cathode, cosmic rays, radioactive sources, photoelectric emission, etc), producing positive ions and additional electrons. When this discharge phase is unable to sustain itself, it is characterized as 'non-self-sustained discharge'.

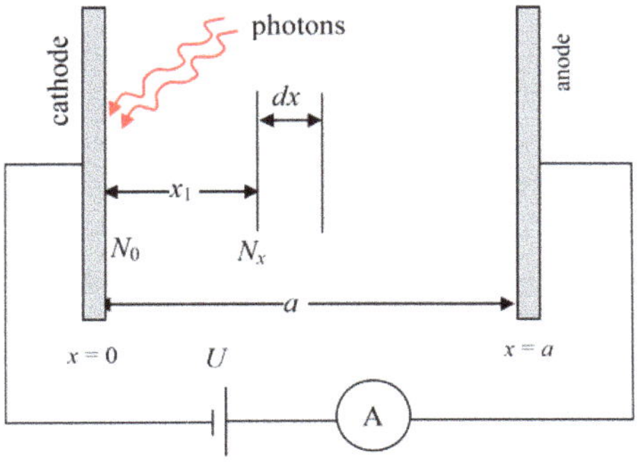

Figure 2.2. Arrangement for the study of a Townsend's discharge process in gases.

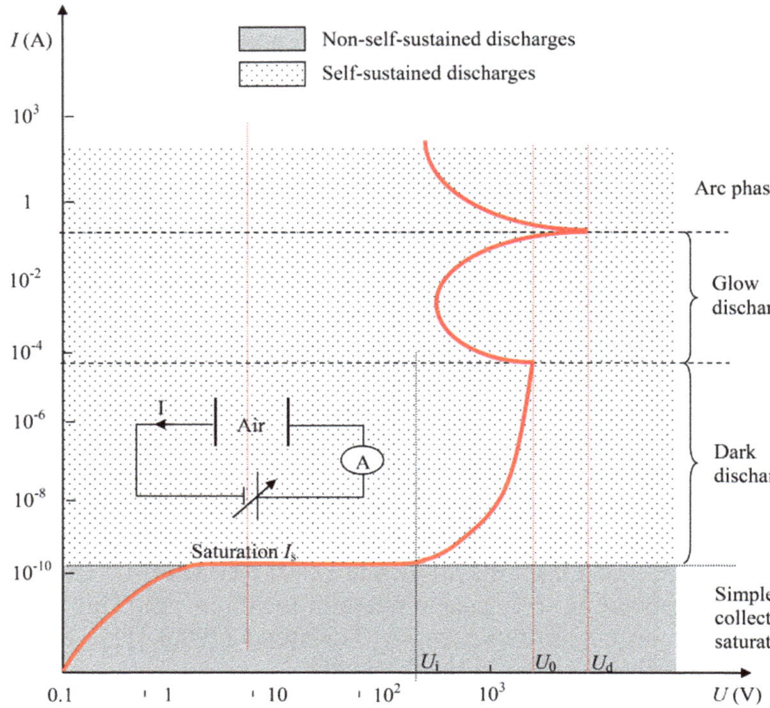

Figure 2.3. Voltage–current (U–I) characteristics of a small air gap under uniform field. $I_s \approx 10^{-10}$ Acm^{-2}; U_i: ionization potential of the gas; and U_d: glow discharge inception voltage.

2.5.2 Non-self-sustaining discharge with electronic multiplication

When slowly increasing the applied voltage (U), the discharge current (I) increases linearly up to a saturation level (I_s) that depends on the gap geometry, such that

$$I_s = 2 \times 10^{-13}\, aS, \tag{2.30}$$

a being the distance between the electrodes (in cm) and S the surface of the electrode (in cm^2). When increasing the applied voltage beyond the ionization potential (U_i) the current starts increasing rapidly. At this stage, the discharge still depends on some external sources to provide ionization. The rapid increase in the current, results from electronic multiplication. Secondary ionization processes, by which secondary electrons are produced, are the ones which may sustain a discharge current. As a result of these ionization processes, positive ions and new electrons are generated. Each newly generated electron contributes, during its drift to the anode, to a real electronic avalanche by the same ionization process (figure 2.4).

The higher the applied electric field, the greater the number of collisions by ionization will be. Let α be the average number per unit length of the ionizing collisions of an electron travelling in the direction of the field; α is called *Townsend's first ionization coefficient*. In addition, let N_0 be the number of electrons emitted by the cathode and N_x the number of electrons at any distance (x) from the cathode.

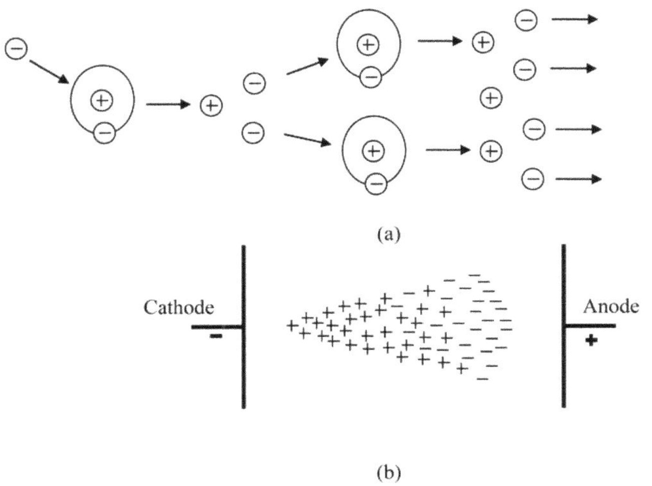

Figure 2.4. Electronic avalanche. (a) Initiation and (b) evolution over time.

The increase of this number, under the influence of ionizing shocks on a further distance dx, assuming α is a constant, is

$$dN_x = N_x \alpha \, dx. \tag{2.31}$$

Then, the number of electrons reaching the anode ($x = d$) is

$$N_x = N_0 \ e^{\alpha a}, \tag{2.32}$$

where $N_0 = N_x(x = 0)$.

As well, the number N_a of electrons reaching the anode is

$$N_a = N_0 \ e^{\alpha a} \tag{2.33}$$

Therefore, the average discharge current flowing in the gap, which is equal to the number of electrons travelling per second, will be

$$I = qN_a = qN_0 \, e^{\alpha a} \tag{2.34}$$

or

$$I_0 = I_0 \ e^{\alpha a}, \tag{2.35}$$

where $I_0 = qN_0$ represents the initial current at the cathode and $q = 1.6 \times 10^{-19}$ C signifies the electron's elementary charge.

It should be noted that the coefficient α depends on the physical condition of the gas and the energy gained by the electrons, i.e. the applied electric field (E). The gas state is characterized by its density (pressure and temperature); later this parameter is practically constant through the ionization process, since the energy transferred to the gas is quite small, given that α depends on gas pressure p and E/p. α has been estimated for several gases by Townsend [12] and other investigators [13, 15–18].

Under the influence of an external electric field (E), the average potential generated during the electron's travel ($n\lambda$) in the cuboid (figure 2.1), is

$$W_{\mathrm{m}} = n\lambda E = \frac{nE}{Ap}. \tag{2.36}$$

Townsend's first ionization coefficient α being proportional to the average value of ionizing collisions per unit length, and in essence to the pressure and energy gained by the electron between two consecutive collisions ($E\lambda$), yields

$$\alpha = pf(E\lambda) = pf\left(\frac{E}{Ap}\right) \tag{2.37}$$

or

$$\frac{\alpha}{p} = f\left(\frac{(E)}{Ap}\right), \tag{2.38}$$

where f depicts an unknown function.

No unique analytic function has been found to describe f analytically. A typical form of this function is shown in figure 2.5. However, an analytical relationship has been proposed by Townsend [9, 11, 12]

$$\frac{\alpha}{p} = A\exp\left(\frac{-Bp}{E}\right). \tag{2.39}$$

A and B are constants characterizing the considered gas. This empirically determined relationship is valid for a given gas in well-defined intervals of E/p. For instance, for air, $A = 15$ and $B = 365$ in the following interval: $100 < E/p < 800$ Vm^{-1} bar^{-1} .

Moreover, equation (2.35) can be rewritten as

$$\ln I = \ln I_0 + \alpha a. \tag{2.40}$$

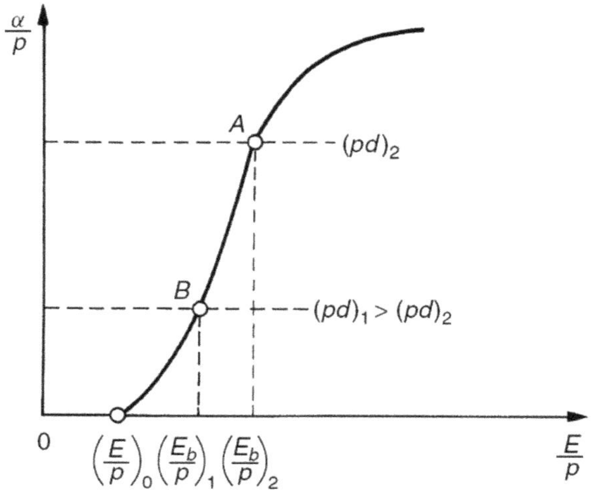

Figure 2.5. Theoretical curve (α/p) as functions of (E/p). Reprinted from [3]. © 1970, with permission from Elsevier.

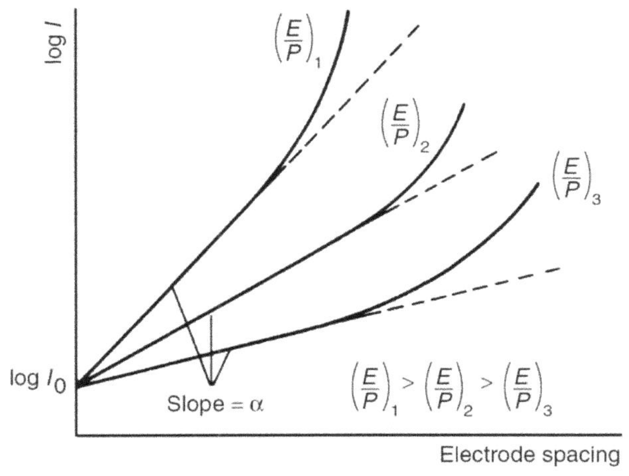

Figure 2.6. ln(I) as function of the gap distance a. Reprinted from [3]. © 1970, with permission from Elsevier.

The diagram ln(I) as a function of the gap distances (a) is theoretically linear with a slope α (figure 2.6). This variation was experimentally verified by A Von Engel for low values of E/p and a, while considering α to be constant [18]. For large values of E/p, other phenomena become involved.

2.6 Self-sustaining discharge

If the number of free electrons on a path of length a (electrode gap), produced by an electron is $e^{\alpha a}$, the number of positive ions simultaneously produced will be $(e^{\alpha a} - 1)$. The total number of electrons emitted by the cathode (N_c) moving to the anode will produce $(N_c(e^{\alpha a} - 1))$ positive ions. While bombarding the cathode, these ions produce $N_c(e^{\alpha a} - 1)\gamma$ additional electrons; γ is the so-called *Townsend's secondary ionization coefficient*. This parameter represents the ratio of the number of electrons emitted from the cathode to the number of positive ions bombarding the cathode.

This Townsend's secondary ionization coefficient γ depends on:

- the type of cathode (i.e. the metal constituting the cathode);
- the type of gas (table 2.3); and
- similarly to (α/p), it also depends on the reduced field (E/p).

The total number of electrons near the cathode is then

$$N_c = N_0 + N_c\gamma(e^{\alpha a} - 1). \tag{2.41}$$

In other words

$$N_c = N_0 \frac{1}{1 - \gamma(e^{\alpha a} - 1)}. \tag{2.42}$$

When moving toward the anode, the N_c electrons take part in ionizing collisions generating additional N_a free electrons in the vicinity of the anode,

$$N_a = N_c e^{\alpha a} = N_0 \frac{e^{\alpha a}}{1 - \gamma(e^{\alpha a} - 1)} \tag{2.43}$$

Table 2.3. Approximate values of the coefficient γ for different gases and the metallic nature of the cathode.

	Gas					
Electrode material	He	Ne	A	H_2	N_2	Air
Al	0.021	0.053	0.12	0.095	0.1	0.035
Ba	0.100	–	0.14	–	0.14	–
Mg	0.031	0.11	0.077	0.125	0.089	0.038
Cu	–	0.02	0.058	0.05	0.066	0.025
Ni	0.015	0.031	0.058	0.053	0.077	0.036
Pt	0.01	0.023	0.058	0.02	0.059	0.017
K	0.17	0.22	0.22	0.22	0.12	0.077
Hg	0.02	–	–	0.008	–	–
C	–	–	–	0.014	–	–
W	–	0.045	–	–	–	–
Fe	0.015	0.022	0.058	0.061	0.059	0.02

Combining equations (32) and (42), the discharge current can be deduced as follows:

$$I = I_0 \frac{e^{\alpha a}}{1 - \gamma(e^{\alpha a} - 1)}. \qquad (2.44)$$

It appears from this relationship that the discharge current I increases very rapidly at the rate of $((1 - (e^{\alpha a} - 1)\gamma)$. According to Townsend's theory, breakdown occurs when the discharge current tends to infinity, that is for $(e^{\alpha a} - 1)\gamma = 1$. Consequently, the corresponding voltage is defined as the breakdown criterion. Note that other phenomena, such as space charge accumulation and electron attachment processes, are neglected.

The condition of a self-sustaining discharge, when neglecting the influence of external sources (ionizing agents), is, e.g.

$$\gamma(e^{\alpha a} - 1) \geqslant 1 \qquad (2.45)$$

or

$$\alpha a \geqslant \ln\left(1 + \frac{1}{\gamma}\right). \qquad (2.46)$$

In such a case, the discharge becomes 'autonomous' or 'self-sustained'. Secondary ionization processes produce secondary electrons beyond the phenomena which sustain the discharge after its initiation due to ionization by collisions and photo-ionization.

The glow discharge inception voltage (U_d) can, therefore, be determined from equation (2.46)

$$\alpha a = \ln\left(1 + \frac{1}{\gamma}\right), \qquad (2.47)$$

assuming that the self-sustaining condition takes place when the applied voltage $U = U_d$ (with $U = Ea$ in uniform field geometry). By combining equations (2.39) and (2.47), the condition for self-sustained discharge at a given temperature can be written as

$$U_d = \frac{B\, p\, a}{\ln\left(A\, p\, a / \ln\left(1 + (1/\gamma)\right)\right)}. \tag{2.48}$$

If the applied field is not affected by space charges, the self-sustained discharge inception voltage will depend only on (pa), where p represents the gas pressure and a the gap distance,

$$U_d = f(p\, a). \tag{2.49}$$

This relationship represents the so-called *Paschen's law*, experimentally known as *Paschen's curves.* Figure 2.7 gives the characteristics of air.

All these curves have the same shape depicting a minimum V_{min}, corresponding to a specific value of $(pa)_{min}$. At large values of pd, the breakdown voltage V increases—this corresponds to high-pressure insulation. At low values of pd, the breakdown voltage V goes to infinity—this corresponds to vacuum insulation; in between, there is a minimum voltage. For instance, this minimum in air corresponds to 350 V for $(pa)_{min}$ = 66.7 Pa cm. This equation means that below 350 V, breakdown could not take place in air for a gap distance of 1 cm. Table 2.4 summarizes some values of V_{min} for various gases, as a function of $(pa)_{min}$, for different types of gases and the material of the cathode.

Equation (2.49) assumes that the temperature T (in kelvins) is constant. However, when taking into account the temperature T, it yields

$$U_d = f(p\, \delta) \tag{2.50}$$

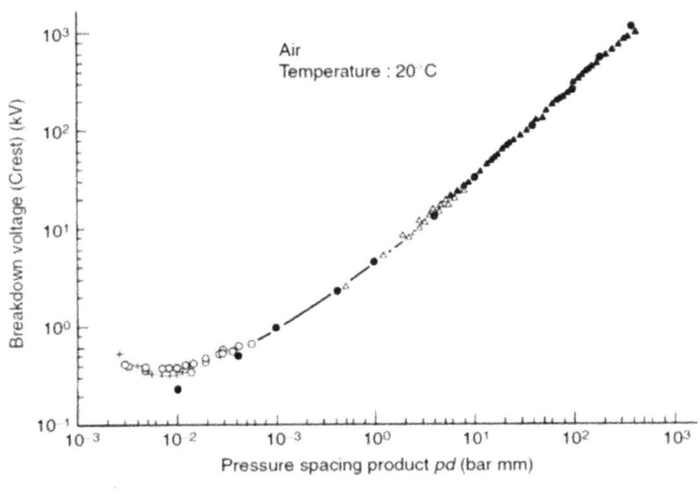

Figure 2.7. Experimental Paschen's curves $U = f(p,a)$ for air. Reprinted from [3]. © 1970, with permission from Elsevier. The dots correspond to experimental measurements issued from several international standards and laboratories.

Table 2.4. The Paschen's minima for different gases.

Gas	Cathode	V_{min}(volts)	$(pa)_{min}$(torr cm)
He	Fe	150	2.5
Ne	Fe	244	3
Air	Fe	330	0.57
A	Fe	265	1.5
N_2	Fe	275	0.75
O_2	Fe	450	0.7
Na	Fe	335	0.04
Hg	Fe	520	2
Hg	W	425	1.8
Hg	Hg	330	–
H_2	Pt	295	1.25

where δ represents the gas density,

$$\delta = \text{const.}\ \frac{P}{T}. \tag{2.51}$$

Paschen's law simply means that for a given electrode gap, due to the limited number of particles at low pressure, the ionizing avalanche is limited which leads to a high breakdown voltage. Conversely, at higher pressure, the large density of particles affects the mean free path (λ); the distance between two collisions is thus reduced. Recall that the ionization process arises from high energy impacts between electrons and atoms. During their course, electrons do not gain enough energy to promote ionization by collision and, consequently, the avalanche phenomenon by ionization. Between both situations, an optimum corresponding to Paschen's minimum is observed.

Townsend's discharge theory may be applied to slightly non-uniform electrode configurations, such as two concentric cylindrical electrodes or two conducting sphere arrangements in which the gap is too small with respect to the radius of the cylinder/spheres.

However, since Townsend's first ionization coefficient (α) depends on the applied electric field (E), which depends on the position x within the gap, the condition of a self-sustained discharge (equation 2.46) modifies to

$$\gamma\left(\exp \int_0^a \alpha \mathrm{d}x - 1\right) \geqslant 1, \tag{2.52}$$

or simply

$$\int_0^a \alpha \mathrm{d}x \geqslant k, \tag{2.53}$$

where $k = \ln(1 + \frac{1}{\gamma})$.

Under normal conditions of temperature and pressure, k is about 20 for air. The presence of space charge and the geometry of the electrodes renders the application of the self-sustained discharge condition difficult (depicted by (2.52)) in the case of non-uniform fields. The electric field corresponding to U_d may not be large enough to sustain the ionization processes by collision. In such a case, the voltage required to break down the gap (U_b) should be higher than U_d. U_b means *breakdown* voltage.

2.7 Limits of Townsend's theory

Electron emission by photo-ionization at the cathode plays an important role in the inception of the self-sustained discharge in Townsend's theory. However, this theory is not able to explain the following:

- There have been experimentally recorded discharge durations of 10^{-9}–10^{-7} s for gap distance a, higher than 1 cm and at atmospheric pressure. Such durations are lower than those required to initiate secondary effects at the cathode (the time necessary for positive ions to reach the cathode, varying between 10^{-5} and 10^{-4} s), says the condition required for the passage from non-self-sustained discharge to self-sustained discharge.
- The inception voltage does not depend on the material of the cathode, as predicted by Townsend discharge theory. According to Townsend, this dependency is related to the coefficient (γ).
- From experimental investigations, discharges were found to be filamentary and irregular [1].

2.8 Streamer–leader theory

The failure of Townsend's theory to explain all the above phenomena led Raether, Meek, and Loeb, to propose an alternative, the *streamer–leader theory*, around 1940 [1]. In their theory, photo-ionization resulting from excitation phenomena and recombination in the gap as well as space charge build-up are the fundamental processes in the development of discharge. According to this theory, the discharge propagates in the form of a conducting channel, called a *streamer* (plasma channel).

The first stage of discharge evolution in channel-like structures corresponds to the initiation of an avalanche, called the *primary avalanche*. This can later be initiated, for instance, by collision ionization once a critical value of the applied electric field is reached at any point of the electrode gap. The growth of charge carriers within an avalanche in a uniform field is described by equation (2.33), provided that field distortion due to ion-based space charge is negligible. When studying the effect of space charge on avalanche growth, Raether noticed that the growth of the avalanche weakened when electronic charge concentration was between 10^6 and 10^8 [2]. However, when the charge concentration increased beyond 10^8, the avalanche current was followed by a steep rise in the current between the electrodes, which led to the breakdown of the gap.

Avalanche growth is accompanied by the generation of space charges. Indeed, in an avalanche electrons build up a negative cloud propagating rapidly towards the anode, while positive ions remain practically static (recalling that electron mobility is about 100–1000 times higher than that of ions).

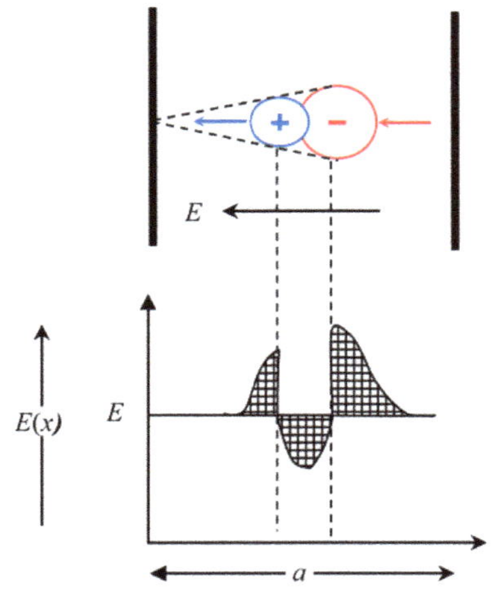

Figure 2.8. A sketch of the field distortion in the gap due to space charge growth.

This renders the field heterogeneous, and its intensity increases near the poles of the avalanche and in particular in the vicinity of the positive pole (constituted by positive ions) which behaves as a conductive point/rod.

The field on the sides of the avalanche decreases. Once the space charge within the primary avalanche reaches a critical size of approximately 10^8 electrons, ionization by collisions at the head and the tail of the avalanche are intensified. The slow growth at low charge concentrations and fast growth at high charge concentrations are attributed to the distortion—by the space charge—of the originally applied uniform field (E). For the sake of simplicity, the space charge at the head of the avalanche can be assumed to be a spherical volume containing negative charge at its top. Figure 2.8 shows the electric field around the avalanche as it progresses along the gap and the resulting modification to the applied field.

Camera recordings of avalanche development have shown that, under certain conditions, the space charge developed in an avalanche can transform the avalanche into streamers which speed up the development of the breakdown process [2].

It can be observed that the electric field is enhanced at the top of the avalanche. Furthermore, at the bottom of the avalanche, the field that exists between electrons and ions reduces the applied field (E) while the field between the cathode and the positive ions is enhanced. Space charge fields, therefore, play an important role in the growth of avalanches in corona and spark discharges. Field distortion becomes noticeable once a charge carrier's concentration reaches around 10^6. At such a level of concentration, as the distortion is significant only in the vicinity of the top of the avalanche, its effect is still negligible. For continuous propagation, this avalanche requires a critical size. It is generally assumed that transformation from an avalanche to a streamer generally occurs when the charge within the avalanche head reaches a critical value of 10^8 electrons. When the charge density in the avalanche reaches this critical value, the space charge filled field and the applied field will have the same magnitude, leading to the initiation of a streamer. The combination of the applied

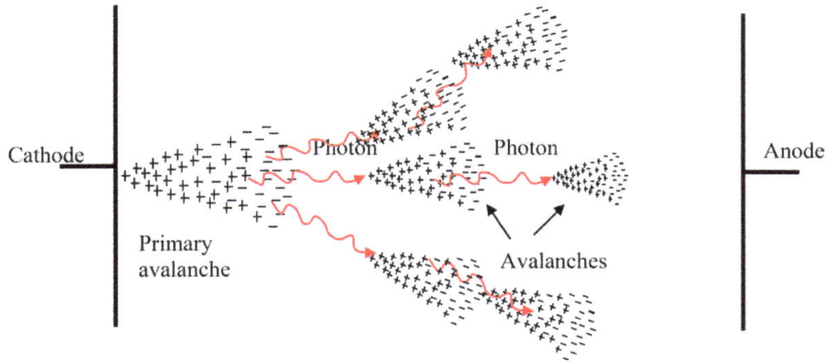

Figure 2.9. A sketch of streamer discharge development by photo-ionizing avalanches.

field and the space charge field causes intense ionization and excitation of the gas particles in front of the avalanche. Instantaneous recombination between positive ions and electrons releases photons, which, in turn, produce secondary electrons by photo-ionization (equations 2.8 and/or 2.9). Excited nitrogen molecules can emit photons with energies higher than 13 eV and thus are likely to ionize oxygen molecules with an ionization potential of 12.2 eV. The newly produced free electrons, accelerated by the electric field, may then initiate new avalanches by collisional ionization, provided the local electric field is high enough (figure 2.9). These avalanches, called *secondary avalanches*, contribute to the development of the self-sustaining discharge. Since photons travel with the velocity of light, photo-ionization processes rapidly increase the development of conduction channels across the gap.

The increase of charge carriers will take place in the field and in the preceding avalanche axis. Each avalanche develops rapidly. While electrons are moving towards the anode, this 'chain' seems to move towards the cathode. The growth of this chain is so fast due to avalanche transplantation by radiation instead of by the migration of electrons and ions. Ionization, therefore, takes place without the influence of the cathode (figure 2.10).

The discharge takes the form of a channel along the electrode axis; its growth depends on the field distribution.

Moreover, during elastic collisions with gas particles, part of the kinetic energy of the charge carriers is later transferred to the gas, increasing its temperature (T_g). As long as T_g remains below the thermal ionization temperature (T_{th}) this increase will not affect the discharge propagation. This stage of development is called a *streamer*. The external field allowing sustained growth is lower than that required for collisional ionization. For example, in air, the average electric field along a streamer propagating from the anode is roughly 5 kV cm^{-1}, while it is 15 kV cm^{-1} when propagating from the cathode. Note that the critical field for corona onset (under normal temperature, pressure, and humidity conditions) is about 30 kV cm^{-1}.

The negative streamer (i.e. an air gap subjected to negative voltage under a divergent field) can be described by a similar mechanism (figure 2.11).

In this case, photons are assumed to create new electrons. However, the triggering of a secondary avalanche must occur in front of the head of the first avalanche,

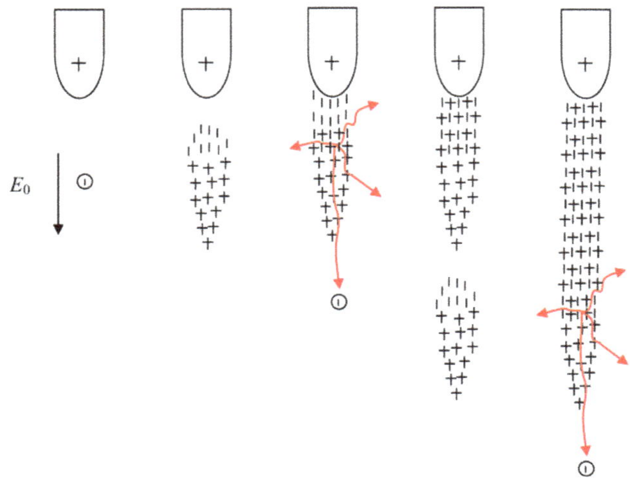

Figure 2.10. A schematic representation of the positive streamer processes.

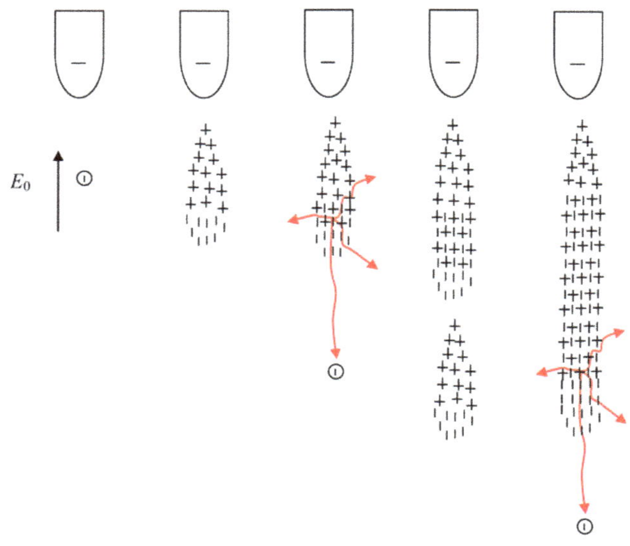

Figure 2.11. A schematic representation of the negative streamer processes.

where the field is reinforced. Consequently, the negative streamer consists essentially of negative charges apparently drifting into the gap. Furthermore, the evolution of the positive space charge that is left behind the first avalanche must be examined. In spite of the high electric field between the positive tail of the first avalanche and the cathode, the positive ions can be assumed to be motionless during the propagation of the negative streamer, due to their low mobility.

Figures 2.12 and 2.13 represent the different stages of channel-like discharge development. The streamer-like discharges, when approaching the opposite electrode,

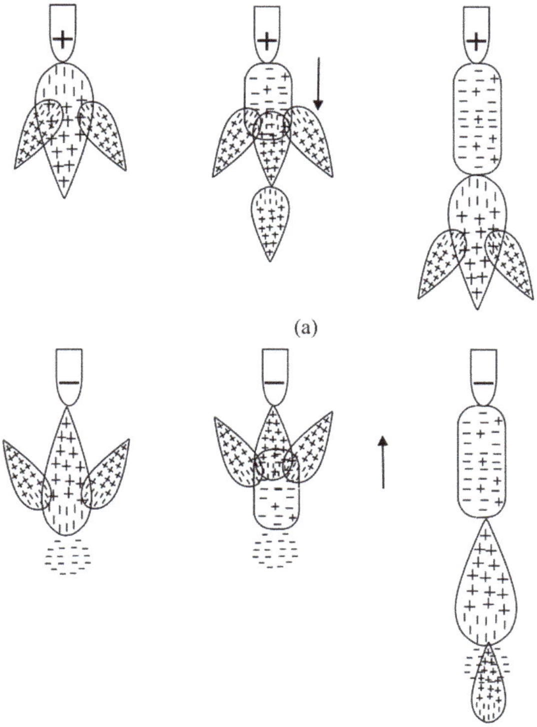

(a)

(b)

Figure 2.12. Streamer discharge growth in a gas under divergent electric field geometry: (a) positive electrode and (b) negative electrode.

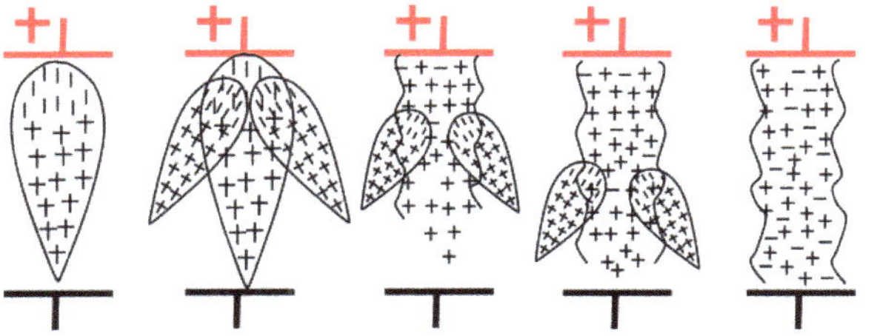

Figure 2.13. Streamer discharge growth in a gas under uniform electric geometry.

significantly increase the field at their head. Furthermore, photo-emission may take place, which leads to an increase in the number of free charges. When the streamer short-circuits the gap, a complete discharge then takes place. The sudden increase in free charge carriers in the discharge channel is accompanied by an increase in the gas temperature up to a value that conditions an intensive thermal ionization ($T_g > T_{th}$). A glow discharge that short-circuits the gap then takes place. Its transition to an arc depends only on the applied voltage characteristics.

For large air gaps (>1 m for air at atmospheric pressure), more complex phenomena are involved. Streamer discharges develop into a *leader*, which is the subject of the next chapter.

References

[1] Loeb L B and Meek J M 1940 *J. Appl. Phys.* **11** 438

[2] Raether H 1964 *Electron Avalanches and Breakdown in Gases* (London: Butterworth) p 390

[3] Kuffel E and Abdullah M 1970 *High Voltage Engineering* (Library of Congress Catalog Card No 72-85501) 1st edn (Amsterdam: Elsevier) pp 322, 338

[4] Meek J M and Craggs J D 1978 *Electrical Breakdown of Gases* (New York: Wiley)

[5] Kind D 1985 *High Voltage Insulation Technology, Textbook for Electrical Engineers* (Weisbaden: Vieweg+Teubner)

[6] Raizer Y 1991 *Gas Discharge Physics* (Berlin: Springer)

[7] Braithwaite N St 2000 *Plasma Sources Sci. Technol.* **9** 517–27

[8] Bazelyan E M and Raizer Yu P 1998 *Spark Discharge* (Boca Raton, FL: CRC)

[9] Townsend J S and Tizard H T 1913 *Proc. R. Soc. Lond. A* **88** 336

[10] Jackson J D 1999 *Classical Electrodynamics* 3rd edn (New York: Wiley)

[11] Townsend J S 1900 *Nature* **62** 340

[12] Townsend J S 1914 *Electricity in Gases* (Oxford: Oxford University Press)

[13] Sanders F H 1932 *Phys. Rev.* **41** 667

[14] Fisher A 1969 *Electra* **10** 65

[15] Marsch K 1932 *Arch. Elektrotech.* **26** 589

[16] Loeb L B 1939 *Fundamental Process of Electrical Discharge in Gases* (New York: Wiley)

[17] Geballe R and Harrison M A 1952 *Phys. Rev.* **85** 372

[18] Von Engel A 1994 *Ionized Gases* (College Park, MD: American Institute of Physics)

IOP Publishing

Discharge in Long Air Gaps
Modelling and applications
A Beroual and I Fofana

Chapter 3

The positive discharge in long air gaps

3.1 Introduction

Positive polarity is of particular importance for transmission systems and/or hardware design. It is considered to be the most severe polarity since the positive breakdown voltage of air gaps is lower than the negative one. Thus, the calculation method for the selection of the phase-to-ground and phase-to-phase insulation withstanding voltages for power network equipment is based on this polarity. The electrode geometry (i.e. gap spacing and electrode shape) is another parameter that plays an important role in the breakdown process, in the sense that it determines the magnitude, direction, and spatial variation of the electric field between the electrodes. Since HV industrial problems are usually related to dielectric problems, the geometry providing the lowest performance in terms of dielectric strength should be considered. From the dielectric point of view, needle/rod–plane configurations represent the most critical situation for an air gap. This type of geometry has some advantages, namely:

- The axis-symmetry with regard to the gap axis allows one to determine the field distribution using cylindrical coordinates.
- This geometry can be simply characterized with only two parameters: the electrode (needle/rod) radius R_p and the electrode gap spacing D;
- A geometry with only one active electrode simplifies investigation since the discharge involves only one electrode during most of its development. Indeed, the cathode (grounded plane) is only involved in the last step (final jump) of the discharge;
- The needle/rod–plane gap geometry subjected to a positive impulse voltage provides the lowest dielectric strength. Based on knowledge of the breakdown strength of needle/rod–plane gap configurations, air gap clearances for industrial applications are determined based on knowing the gap factor [1].

The initiation and propagation of the positive leader in non-uniform fields have been thoroughly investigated experimentally as well as theoretically by many

doi:10.1088/978-0-7503-1236-3ch3

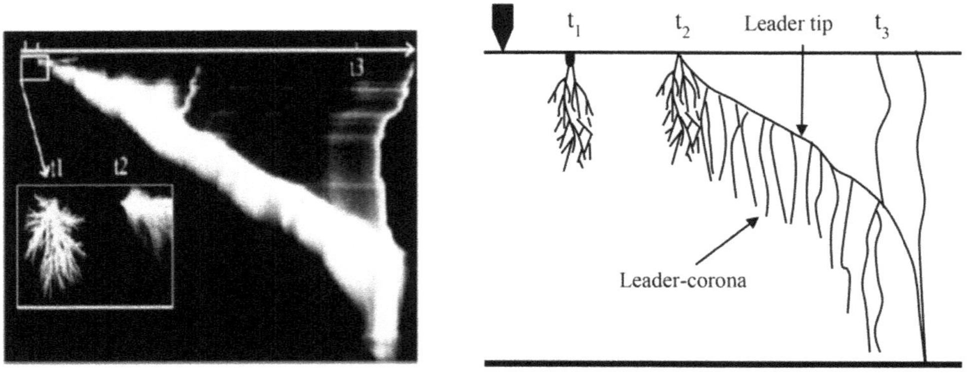

Figure 3.1. Streak photograph and sketch showing the development of a laboratory positive discharge (10 m gap, t_m = 500 μs, V_{max} = 2.5 MV) [11].

researchers [1–15]. Basically, the breakdown of an air gap under an impulse voltage is not an instantaneous phenomenon. Figure 3.1 depicts successive images observed by means of an image converter (digital camera) working in streak mode when a long air gap is subjected to a positive impulse wave.

In this chapter the fundamental processes of the subsequent phases of long air gap discharges are presented, starting from the inception and development of the first corona, to the formation and propagation of the leader channel, and up to the final jump (return stroke).

3.2 Air gap breakdown process under an impulse voltage

A typical development of a positive discharge in a point–plane air gap subjected to a positive impulse voltage is shown in figure 3.2 [2]. There are three main steps leading to breakdown:

- the first corona's inception with or without dark periods;
- leader formation and development; and
- the final jump and arc phase.

3.2.1 First corona inception

The inception of the spark in long air gaps under a highly non-uniform electric field is preceded by a corona discharge in the form of hair bright filaments (streamers) called streamer corona, at time t_i. Corona inception is known to be a statistical event. Initial electrons are needed to start the process as described in the previous chapter. Such electrons may either arise from cosmic radiation, natural ionization sources, residual charge from a previous discharge, or as a consequence of field emission from surface asperities.

An average of 7–20 electrons may be released per cubic centimetre per second by natural ionizing radiation). These electrons are generally attached to oxygen molecules

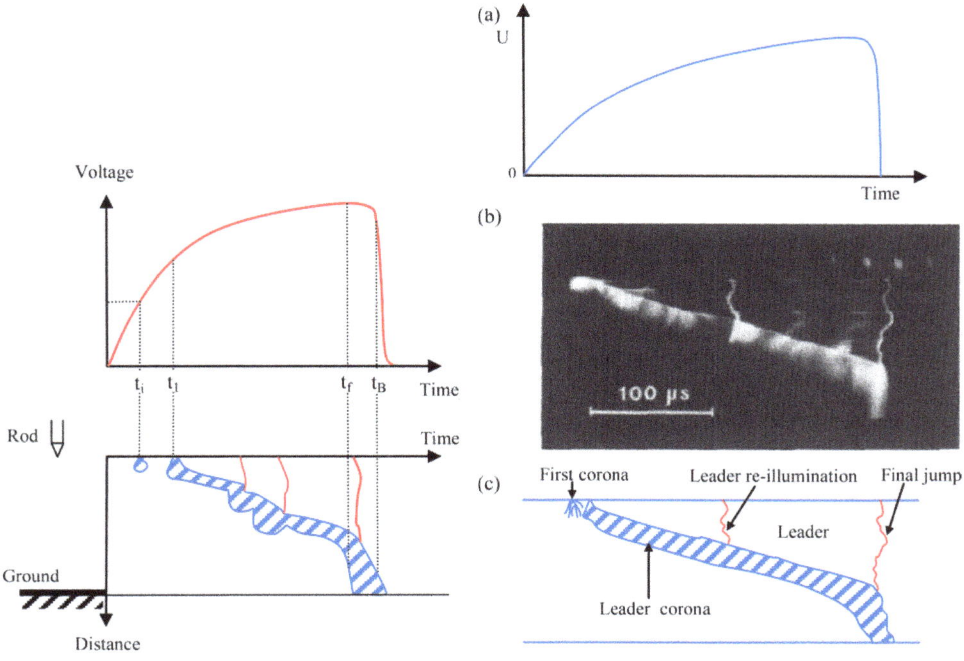

Figure 3.2. Diagram showing the successive phenomena observed using an image converter working in streak mode when a point–plane air gap of 8 m is subjected to a positive impulse voltage 500/1000 μs [2]. (a) Voltage waveform applied to the gap. (b) Streak photograph of a long spark developing in a positive point–plane gap. (c) Schematic representation of the discharge of (b).

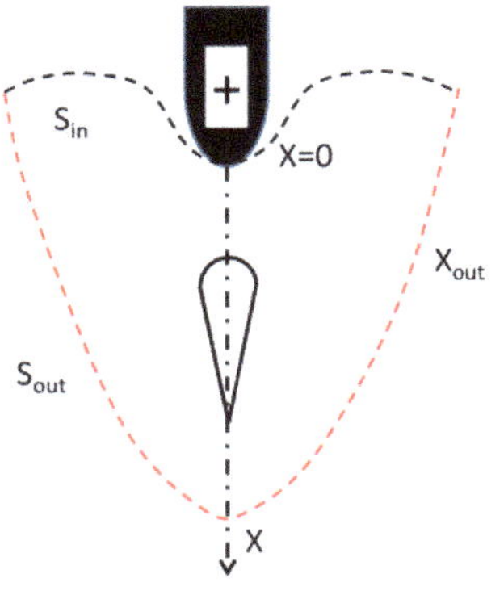

Figure 3.3. Representation of the critical volume in a positive discharge sequence.

and turn into negative ions. Under positive high electric stress, negative ions may drift towards high field regions and produce new free electrons through collisions [4, 5].

To initiate critical size avalanches, one of these electrons must be suitably placed, i.e. inside the so-called 'critical volume' (figure 3.3). The critical volume (or active region) is the volume of gas surrounding the HV electrode within which critical size avalanches can transform into streamers (see section 2.8).

This critical volume is defined by the following two conditions:

- The electric field must be high enough (higher than the critical field, E_c) and the electron must be located in the region where $\alpha - \eta > 0$ (where α and η are the ionization and attachment coefficients of the gas, respectively).
- The avalanche initiated by the initial electron must develop over a sufficient distance in order to transform into a streamer.

From these conditions two boundary surfaces can be defined, S_{int} and S_{ext} of this critical volume:

$$S_{int} : x = 0 \qquad \text{and} \qquad \exp \int_{x_{int}}^{x_0} (\alpha - \eta) dx^3 10^8 \tag{3.1}$$

$$S_{out} : x \neq 0 \qquad \text{and} \qquad E = E_c. \tag{3.2}$$

The electric field (E_c) depends on voltage characteristics, electrode geometry, and atmospheric conditions. Its values vary in the range from 24–30 kV cm^{-1} for typical configurations.

In 1929, Peek [16] developed an empirical relationship, known as *Peek's law* to describe the conditions necessary for corona discharge inception between two conductors (cylindrical wires):

$$\boxed{E_c = E_0 m\delta\left(1 + \frac{k}{\sqrt{\delta r}}\right) \text{kV eff cm}^{-1}}, \tag{3.3}$$

where m is a parameter related to the conductor's surface ($0 < m < 1$). This parameter is affected by roughness, pollution, rain, etc. For instance, $m = 1$ for smooth and very polished cylindrical wires; $m = 0.8$–0.9 for stranded conductors; and $m = 0.3$–0.6 under rainy conditions. r is the radius of the equivalent conductor in centimetres. r can be derived as a function of the number of conductors per phase (bundle conductors), n, the diameter of the conductor (d_c in cm), and the diameter of the bundle (d_f in cm):

$$r = \left[n d_c d_f^{(n-1)}\right]^{1/n}. \tag{3.4}$$

δ is the relative air density. Under normal conditions of pressure and temperature ($p = 1013$ mbar, $T = 293$ K), $\delta = 1$. However, for given pressure p and temperature T values, δ is calculated as:

$$\delta = \frac{0.392\, b}{T} = \frac{2.92\, c}{T}, \tag{3.5}$$

where b is the barometric pressure in mmHg, c the barometric pressure in kPa and T is the temperature in Kelvin.

The parameters E_0 and k depend on the applied voltage characteristics (waveform and polarity). Their values can be taken from table 3.1.

When an initial electron is available in the critical volume, an electronic avalanche is initiated provided the conditions specified in equations (3.2) and (3.3)

Table 3.1. E_0 and k as functions of voltage polarity.

	E_0	k
Alternative voltage	21.1 kV rms cm^{-1}	0.3
Positive dc voltage	33.7 kV cm^{-1}	0.24
Negative dc voltage	31 kV cm^{-1}	0.38

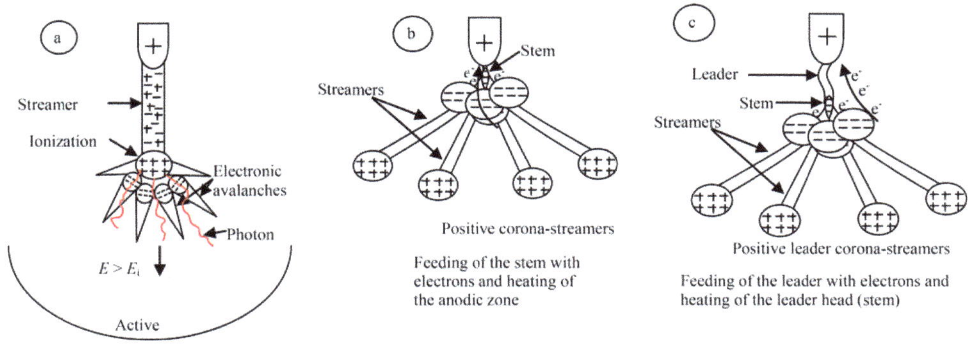

Figure 3.4. A sketch of the sequences leading to corona leader formation: (a) positive streamer formation; (b) feeding of the corona root with electrons and anodic stem formation; and (c) feeding of the leader head (stem) with electrons and elongation of the leader.

are fulfilled. The first avalanche that develops (see section 2.7) is followed by secondary avalanches which further transform into streamers (see section 2.8). Ahead of the streamers, new avalanches develop (figure 3.4(a)). This process gives rise to a 'hair' of luminous filaments in conical form (at the HV electrode or rod) known as 'corona-streamers'.

The corona consists of a number of narrow branched channels or 'streamers', which develop from a common root. This corona represents the first visible luminous activity at the HV electrode, after the application of the impulse voltage at T_i defined in figure 3.2. Figure 3.5 depicts static photographs of first positive corona from (a) a 250 mm sphere for an electrode gap of 4 m [3] and (b) a point electrode of 2 mm radius of curvature for an electrode gap of 0.5 m [17]. The mean electric field along the streamers' filaments is about 5×10^5 Vm^{-1}, while their temperature is close to the ambient value. This type of discharge is generally characterized as a 'cold discharge'; the lifetime of this first corona, about 300–400 ns, is too short to involve any heating of gas. The only zone likely to undergo a consequent heating is the root of the first corona.

3.2.2 The dark period

The first corona discharge can be followed (in the case of a divergent electric field) by one or more periods, in between which other corona effects are initiated; this is the so-called dark period during which no luminous process takes place within the gap (figure 3.6). The absence of any ionization phenomena during the period $T_d = T_1 - T_i$ can be explained as follows. At time T_i, the field is reduced below the ionization level E_i by the space charges ΔQ_m injected by the first corona. Indeed, the latter tend to

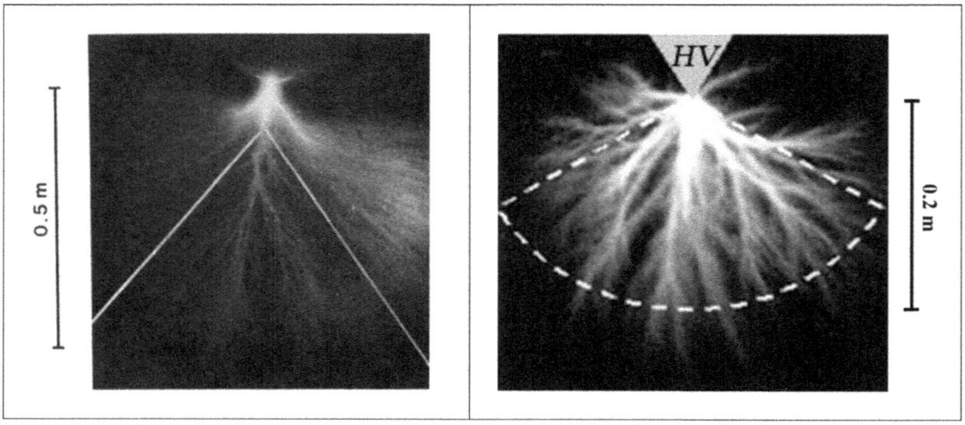

Figure 3.5. Static photographs of first positive corona from (a) a 250 mm sphere for an electrode gap of 4 m [3] (figure reprinted with permission from CIGRE, Positive discharges in long air gaps at Les Renardieres— © 1977) and (b) a point electrode of 2 mm radius of curvature for an electrode gap of 0.5 m [17] (reproduced with permission of Université de Pau et des Pays de l'Adour, France).

quench the phenomenon whereas the geometrical field, which increases in speed according to the applied voltage shape, is the driving element. The electric field is then reduced below E_i at time T_i accordingly (see figure 3.6).

The geometric field will increase again up to E_i (depending on the shape of the applied voltage). The dark period ends at the time t_1 with the inception of a new corona. T_d depends on the first corona inception voltage as depicted in figure 3.7.

It should be emphasized that the existence and duration of the dark period depends on various inter-related parameters [1]. For fast-front wave shapes with a very small crest time T_{cr} (e.g. a lightning impulse) at the inception of the first corona, the inception voltage (U_i) is high, without a dark period, due to the rapid increase in the applied voltage during the corona development.

It seems that the existence and duration of this dark period depends on the shape of the HV electrode (radius of curvature); the smaller the curvature radius of this electrode and the higher T_{cr}, the more marked the phenomenon is. In fact, the absence or the repetition of this phase depends on the relative values of the geometrical field and the space charge field induced by the corona.

3.2.3 The leader

Leader formation

After one or more coronas, if the voltage at the electrode is high enough, the leader channel inception itself takes place at time t_1 (figure 3.1), starting from the common root of the coronas ('stem') as shown in figure 3.4(b).

Indeed, as long as the applied voltage increases, secondary avalanches intensify. In the vicinity of the HV electrode, the density of the streamers is excessively high. While coronas are growing, this zone is crossed by a significant number of electrons collected at the anode (rod electrode). It is obvious that the mechanism of the corona–spark transition must involve heating of air to a thermal ionization

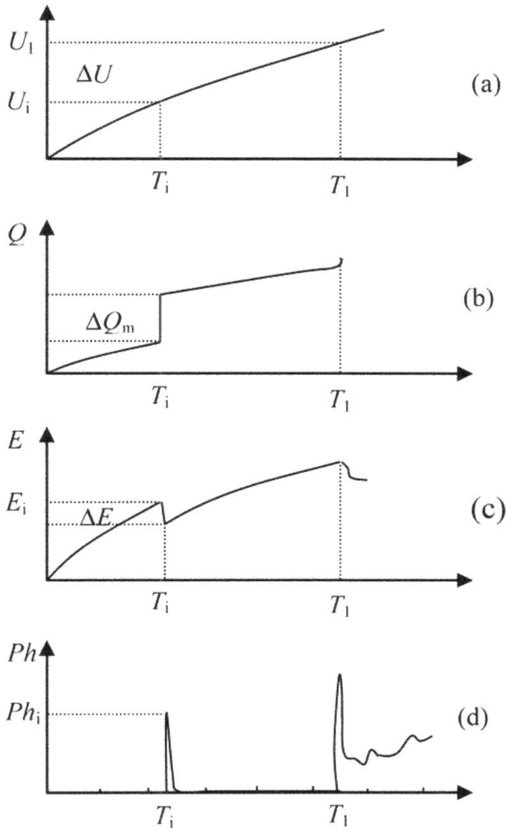

Figure 3.6. Typical oscillograms of phenomena related to first corona [1]: (a) applied voltage; (b) charge injected into the gap; (c) electric field measured at the HV electrode; and (d) emitted light.

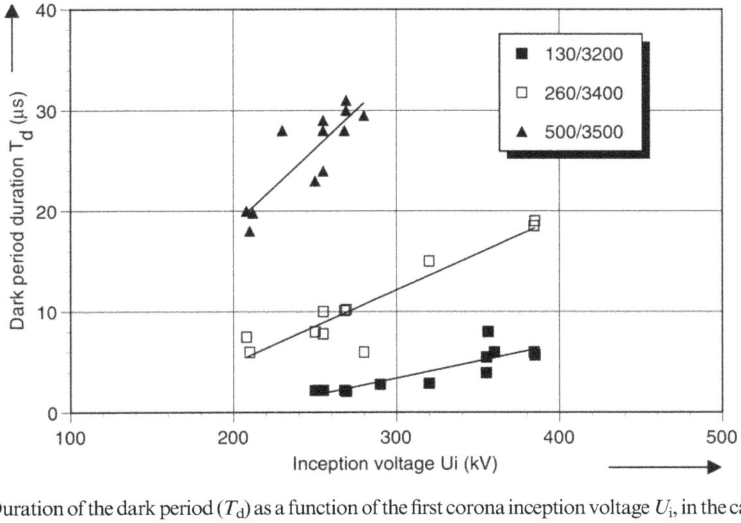

Figure 3.7. Duration of the dark period (T_d) as a function of the first corona inception voltage U_i, in the case of a conical HV electrode, a 5 m air gap distance, and for various voltage waveforms, reproduced from [1] with permission.

temperature. Although each streamer filament taken individually is cold and 'less conducting', the root of their concentration, where the electrons drift, undergoes a local heating of gas. This increase in temperature causes detachment of electrons (attached as negative ions) and/or a reduction in attachment processes. This stage is followed by an increase in conductivity that results from the transition of an attaching medium to a non-attaching one.

This phenomenon is crucial since it induces a local modification of the gas structure that affects the further development of the discharge. If the intensity and duration of the corona are appropriate, this zone is reached at t_2 (see figure 3.1), the physical properties—not currently quantified in detail—transform the corona into a 'stem'. This stem transforms into (or has already become) a leader (figure 3.4(c)).

The leader appears as a weakly luminous channel containing a few negative ions, electrons or positive ions in *non-local thermodynamic equilibrium* (non-LTE), where the energy is transferred from the electric field to the gas by the electrons via collisions.

Local thermodynamic equilibrium

Consider a column of gas (or a mixture of gases) including neutral particles (atoms, molecules) and charged particles (ions, electrons), excited or not. The temperature within the column is set at a value T. Each particle has the same mean kinetic energy. If T is high enough, ionization, recombination, excitation or de-excitation, photon emission or absorption, etc, may take place. Equilibrium can be achieved for a certain value of temperature T. In such a case, the medium is considered to be in thermodynamic equilibrium.

In the discharge column, the energy is transferred to the column by the applied electric field. Due to their relatively low mobility and large mass, ions are generally less affected by the electric field than other particles (electrons). At the opposite end of the continuum, electrons are accelerated and transfer their kinetic energy to heavy particles through elastic collisions. Strictly speaking, heavy particles (ions) will never have the same temperature as electrons. They will be able to approach the electron temperature without ever reaching it. The temperature of heavy particles may be close to the electron temperature provided that the frequency of elastic collisions between *electrons and heavy particles* is as high as possible. This implies a high electronic density. For all these reasons, thermodynamic equilibrium is never actually achieved in laboratory arc plasmas. However, it is possible to define conditions for its application locally, known as LTE. Figure 3.8 describes the progressive transition to thermodynamic equilibrium with increasing current.

Leader propagation

The applied potential at the HV electrode is almost transferred at the tip of the leader, where new coronas will be able to start. The mean electric field along the streamer filaments is about 5 kV cm^{-1}, while this value ranges from 0.1–1 kV cm^{-1} for the leader channel due to the large concentration of charged particles (10^{17}–10^{18} cm^{-3}) at a few thousand degrees Celsius [3]. Under stable propagation conditions, the leader is fed by a continuous current of a few amperes [1, 3]. The advancement of the conductive channel into the gap keeps the electric field at the front of the leader

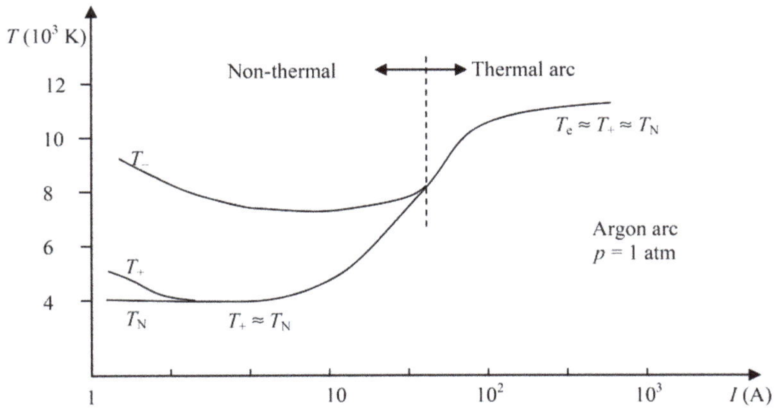

Figure 3.8. Definition of LTE [18]. Reproduced from Arc Physics 1968 Hoyaux © Springer with permission of Springer.

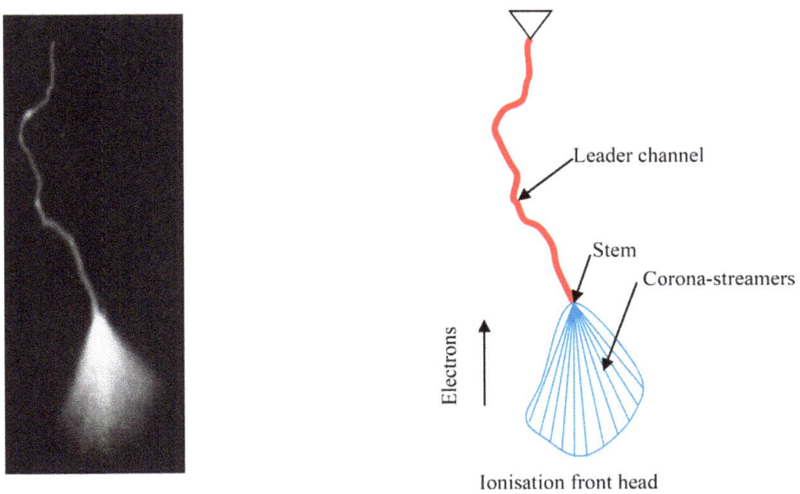

Figure 3.9. Positive discharge structures with an indicator of electron drift direction. Reproduced from [2] with permission of Revue Générale de l'Electricité.

corona sufficiently high to sustain collisions by ionization processes. The discharge structure consists of a coupled propagation of the leader channel simply through a feeding corona discharge (formed of streamers) at its head, also called the 'leader corona' (figure 3.9) [19].

Based on short gap measurements, the streamers that form the corona are essentially constituted of two regions [1–5]:

1. *The streamer head or active region* which is the site of luminous emission and ionization processes. The active region has the following characteristics [5]:
   (i) The streamer head radius is in the range of 10–30 μm, and the net electrical charge density is of the order of 10^{21} m^{-3}.

   (ii) The rotational temperature in this region does not exceed 330 K while the vibrational temperature is raised far above 1000 K; the vibrational energy is then relaxed for long periods because of the long lifetimes of the vibrational state.
   (iii) The electron energy at the streamer head is in the range 5–15 eV, which corresponds to average fields of 10–15 MV m^{-1} [5].
2. *The streamer channel or passive region* where the electrons generated at the tip flow towards the HV electrode and remain attached to the electronegative molecules. The passive region has the following characteristics:
   (i) The streamer channel radius, estimated in short gaps, ranges between 10–30 μm, while the corresponding electron density amounts to 5×10^{19}–10^{21} m^{-3} [5]. It should be emphasized that the electron density may be much lower in large air gaps, due to the attachment process which reduces the free electron population.
   (ii) The current is essentially an electronic conduction one that flows in a resistive regime along all the streamer channel.
   (iii) The dominant electron collision processes are three-body and dissociative attachment.

The 'leader corona' is a diffused corona region, which supplies the current necessary to sustain the leader propagation. This corona that develops in a pre-ionized medium presents some characteristics similar to the first corona. The negative ions generated by the preceding coronas facilitate the production of electrons and decrease the statistical time lag required for subsequent corona inception. These coronas are more ramified than the first ones, but they radiate in the same UV spectrum. On the other hand, the leader radiates in the infrared spectrum, since this discharge is in the first level of thermalization ($T > 1500$ K). At such a temperature, electrons can be detached from a large number of negative ions present in the streamer; electron attachment is compensated later on by their accelerated detachment. The temperature of the leader continues to increase due to the surge in its conductivity. The structure of the leader is physically different from that of the streamers (figure 3.10). Spectral measurements [20, 21] revealed that streamers radiate between 300–400 nm, while the leader emits in the infrared band, i.e. between 600–800 nm. Recall that the visible spectrum for humans ranges between 400–700 nm.

Various optical measurement techniques have been used to study the leader channel [3, 5, 20, 21]. Two camera recordings taken with an image converter in streak mode (figure 3.11) or frame mode (figures 3.12 and 3.13) show a sequential propagation of the discharge. The structure of the leader shows a bright channel and an extensive zone of diffuse glow. Inside the leader channel, thermal ionization is the dominant process. An erratic propagation of the leader channel can be observed in these pictures, which is mainly due to the positive space charge produced by the coronas at the head of the leader. The positive space charge has an antagonistic effect on the development of the leader. In the zone located between the head of the coronas and the leader, the space charge field is opposed to the applied field.

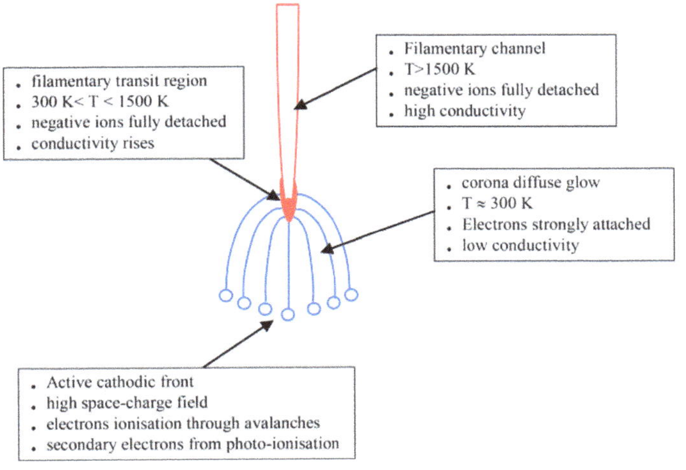

Figure 3.10. The physical structure in the front of a propagating leader discharge [5]. Courtesy of Gallimberti *et al*

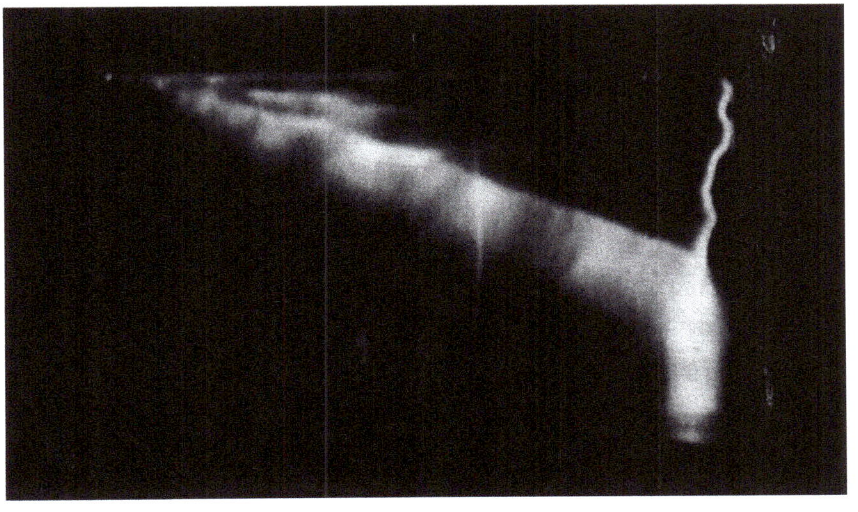

Figure 3.11. Development of a positive discharge recorded by an image converter working in streak mode (10 m point–plane gap, 500/10000 μs wave; the radius of the curvature of point is 1 cm). Reproduced from [2] with permission of Revue Générale de l'Electricité.

This competition plays a fundamental role in the advancement of the discharge. Every new corona avoids the 'fingerprints' left by the preceding one in order to propagate under optimal conditions, as can be observed in the first three images (from the left) of figure 3.12. In figure 3.14, based on Alexandrov's concept depicting this phenomenon, the new corona does not develop in the directional axis of the preceding one.

The new stem ahead of the leader contributes to its lengthening. The flow of electrons resulting from the ongoing corona effect contributes to the temperature

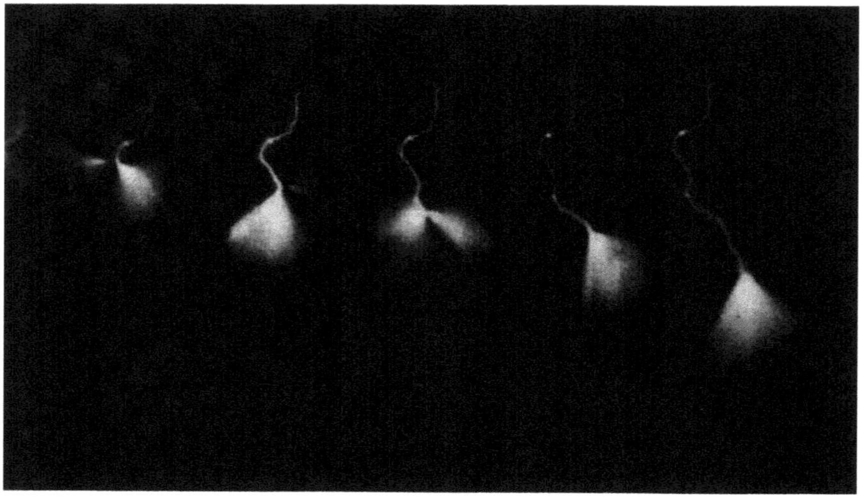

Figure 3.12. Development of a positive discharge recorded by an image converter working in frame mode (10 m point–plane gap, 500/10000 μs wave; the radius of the curvature of point is 1 cm). Reproduced from [2] with permission of Revue Générale de l'Electricité.

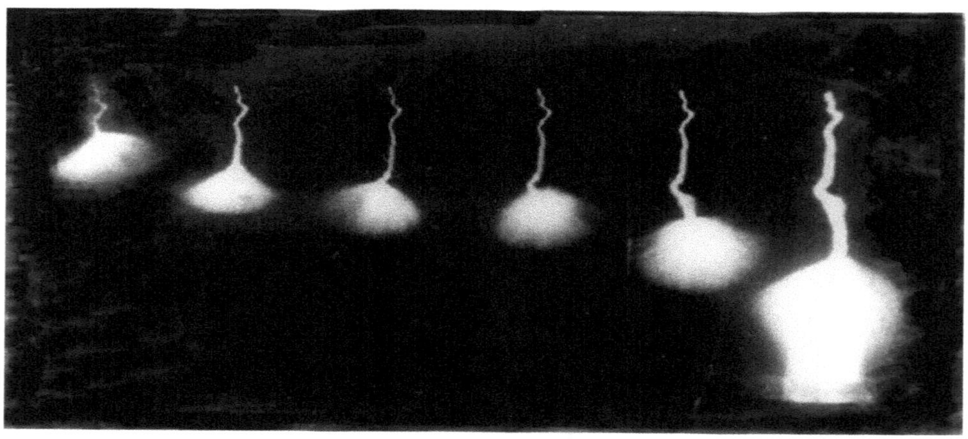

Figure 3.13. A frame photograph of a leader channel connected to the luminous corona's conic structure (10 m cone–plane gap, 500/10000 μs wave) reproduced from [5].

and conductivity of this over time. A re-illumination of the leader may then take place. The intensity of this re-illumination is directly related to electron flow and consequently to the corona ahead of the leader. The propagation of the leader is dependent on the applied voltage $U_c(t)$ growth rate. Each elementary step is directly related to the corona at the head of this leader; the repetition rate of these elementary steps depends on the local field. The leader's ability to propagate through the gap is conditioned by the field at its tip together with the field in the front corona.

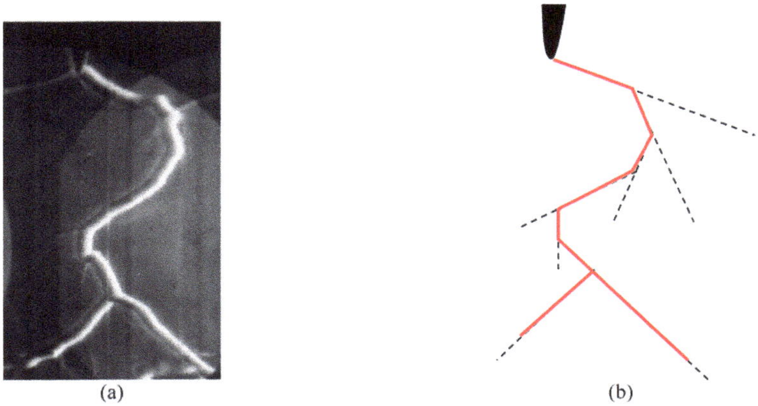

Figure 3.14. A representation of leader development: (a) a strioscopic recording reproduced from [1] and (b) a schematic representation. Only the axis and the successive corona effects are indicated.

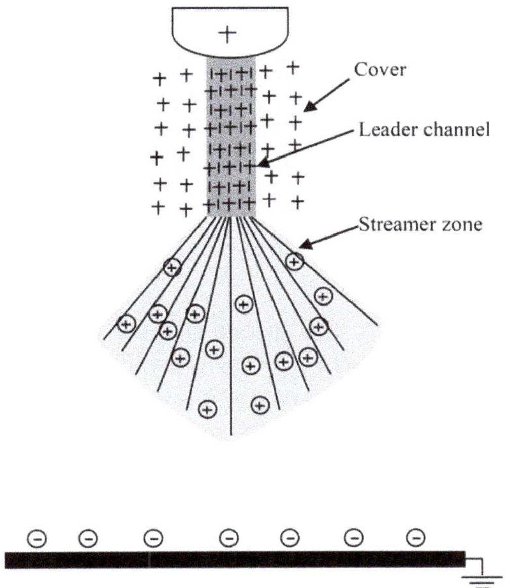

Figure 3.15. Scheme of the positive leader with its cover.

Under constant voltage, the field at the tip of the leader decreases during its elongation. This phenomenon slows down the leader process and may even stop it. The subsequent coronas generated after this period induce a sharp re-illumination of the leader channel. These re-illuminations (or re-strikes) are characteristic of the slow-front impulse voltages [1, 3, 4, 20]. The generation of new corona contributes to the elongation of the leader, i.e. the creation of a hot and conducting channel which extends in the gap towards the cathode with a speed of about 2 cm μs^{-1} [1, 3, 4, 20]. Space charges carried by the corona-streamers are accumulated in the cover of the leader channel (figure 3.15). Indeed, as the leader propagates, the streamer zone also travels through the gap. Negative charges are generated to the grounded electrode through electrostatic influence (figure 3.15).

The initial streamers and their space charge do not move forward; as a result, the elongating leader enters the space filled with charge. The charge cover is not at all rigid, but there is a continuous charge redistribution going on that may be accompanied by ionization events in the radial field at the channel surface. As a result, the current varies along the leader axis.

During the continuous propagation of the discharge through the gap, the corona-streamer–leader system can be represented as depicted in figure 3.15. The diameter of the conducting part of the leader lies between 0.1 mm and 2 mm [3, 22].

The measurement of the leader's physical parameters (temperature, conductivity, electric field, and so on) is difficult to carry out, since sophisticated pieces of equipment are needed. Estimated or indirectly measured values can be found in the literature [1, 19–24].

It is generally acknowledged that the gradient along the leader is about 1 kV cm^{-1} [19–22]. As well, some investigators estimate the temperature of the leader to be between 5000 K and 6000 K [1, 3, 4, 22]. Its conductivity is about 10^{-5} $\Omega^{-1}m^{-1}$, which is likely to vary considerably as a function of the experimental procedures, whereas the diameter of the conducting channel is 0.01–0.2 cm [3, 4, 22].

Under a divergent field, the discharge channel does not necessarily follow the lines of the applied field and may move away from the gap axis. When the leader moves out of range of the gap axis, into a region where the geometrical field is too weak, its development may stop.

3.2.4 The final jump and arc phase

While the leader is propagating, the corona-streamers can reach the cathode (see position 1 in figure 3.16). At this stage, electrons are stripped from the cathode to feed the leader channel. The increase in the degree of ionization might explain the intensive re-illumination of the streamers. The velocity of the leader increases quasi-exponentially

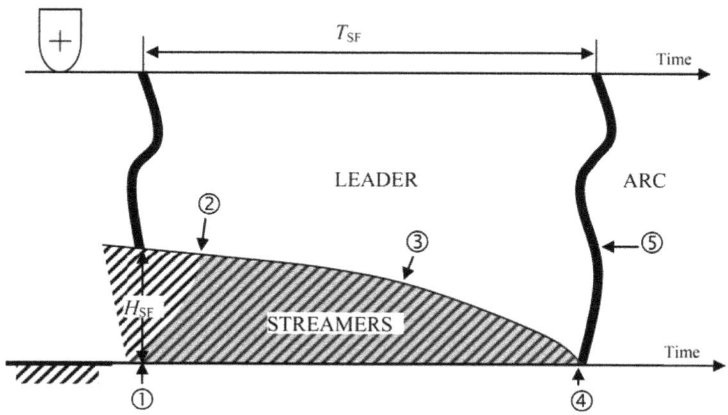

Figure 3.16. Schematic representation of the final jump: 1. Streamers' arrival at the grounded plane. 2. Streamers' re-illumination. 3. Head of the leader. 4. Leader's arrival at the grounded plane. 5. Return stroke.

and reaches a few metres per second. The gap is now ready to be short-circuited: this is the so-called 'final jump'. When the corona leader arrives at the cathode, the last step is completed practically in a straight line by the leader, which thus short-circuits the gap between the electrodes. Thus, the ultimate phase is completed (see position 4 in figure 3.2): this is the return stroke (the main spark stage) responsible for the vivid illumination observed during an electrical discharge into air. From experimental observations, this phase seems to be unaffected by the gap geometry and voltage wave form applied to it. The duration of the final jump is almost constant and is equal to 20 μs [1].

The current flowing in the channel may be several orders of magnitude higher than that of the first stage, sometimes as high as 10^2–10^3 A, even in laboratory spark gaps. The current increase is accompanied by a high energy input, leading to a rapid increase in the channel temperature and conductivity while the voltage drops. This stage is followed by a shock wave, followed by the junction, one microsecond later LTE applies in the channel: this is the arc phase. In practice, it is a transient arc that discharges the capacitances of the HV generator. The gap voltage collapses completely; therefore, particle recombination becomes the dominant process. The subsequent return stroke or arc phase that completes the breakdown process occurs. The current flowing in the channel increases rapidly with a concomitant rise in the degree of ionization.

References

[1] Le Roy G, Gary C, Hutzler B, Lalot J and Dubanton C 1984 *Les Propriétés diélectriques de l'Air et les Très Hautes Tensions* (*Collection de la Direction des Études et Recherches d'Électricité de France vol 51*) (Paris: Eyrolles) p 439

[2] Berger G, Baillot A, Gallet G, Garcia H N, Gary C, Goldman M, Hutzler B, Jouaire J, Leroy G and Simon M 1974 *Rev. Gen. Electr.* **83** 761

[3] Les Renardières Group 1977 *Electra* **53** 31

[4] Gallimberti I 1979 *J. Physique* **40** 193

[5] Gallimberti I, Bacchiega G and Bondiou-Clergerie A and Lalande Ph 2002 *C R Phys.* **3** 1335

[6] Meek J M and Graggs J D 1978 *Electrical Breakdown of Gases* (New York: Wiley)

[7] Raizer Yu P 1997 *Gas Discharge Physics* (New York: Springer)

[8] Kuffel E and Zaengl W S 1984 *High Voltage Engineering Fundamentals* (Oxford: Pergamon)

[9] Von Engel A H 1964 Ionised Gases (Oxford: Oxford University Press)

[10] Naidu M S and Kamaraju V 1996 *High Voltage Engineering* 2nd edn (New York: McGraw-Hill)

[11] Hutzler B and Hutzler D 1982 Modélisation de l'amorçage des grands intervalles d'air *Bull. Dir. Etud. Rech.* **4** 11

[12] Waters R T 1978 Spark breakdown in non-uniform fields *Electrical Breakdown of Gases* (New York: Wiley)

[13] Bondiou A and Gallimberti I 1994 *J. Phys. D: Appl. Phys.* **27** 1252

[14] Fofana I 1996 Modélisation des décharges positives dans les grands intervalles d'air *PhD Thesis* Ecole Centrale de Lyon, University of Lyon

[15] Rizk F A M and Vidal F 2008 *IEEE Transactions on Power Delivery* **23** 296

[16] Peek F W 1929 *Dielectric Phenomena in High Voltage Engineering* (New York: McGraw-Hill)

[17] Gilbert A 1976 private communication see Ortéga P, Heilbronner F, Rühling F, Diaz R and Rodière M, 2005 *J. Phys. D: Appl. Phys.* **38** 2215
[18] Hoyaux M F 1968 *Arc Physics* (New York: Springer)
[19] Domens P 1987 Contribution à l'étude des décharges électriques dans les grands intervalles d'air *PhD Thesis* Université de Pau et des Pays de l'Adour, France
[20] Ortéga P 1992 Comportement diélectrique des grands intervalles d'air soumis à des ondes de tension de polarité positive ou négative *PhD Thesis* Université de Pau et des Pays de l'Adour, France
[21] Rees T 1997 Contribution à l'étude eChéorique des décharges électriques de polarité négative dans l'air à pression atmosphérique *PhD Thesis* Université de Pau et des pays de l'Adour, France
[22] Les Renardières Group 1986 *Proc. IEEE* A **133** 395
[23] Klewe R C, Waters R T and Jones B 1974 Breakdown phenomena in long air gaps under switching impulse conditions *IEEE Summer Meeting* paper 74-CH 0910 PWR
[24] Working Group 07 (Dielectric strength of external insulation under transient voltage) of Study Committee 33 (Overvoltages and Insulation Coordination) 1992 Guidelines for evaluation of the dielectric strength of external insulation CIGRE Brochure No. 72

IOP Publishing

Discharge in Long Air Gaps
Modelling and applications
A Beroual and I Fofana

Chapter 4

The negative discharge in long air gaps

4.1 Introduction

The dielectric behaviour of breakdown in air gaps in a non-uniform electric field distribution under negative polarity is of interest from an engineering point of view, even though positive switching impulses generally provide the lowest dielectric strength. In practice, negative discharges are present in nearly every engineering configuration. Even when not predominant in the breakdown process, they can play a significant role in multi-electrodes and phase-to-phase arrangements. Most stresses caused by lightning in electrical networks relate to negative polarity; this being of special interest for the determination of failure risk in power transmission systems [1–3]. Furthermore, the study of negative discharge can improve our basic knowledge of the lightning discharge mechanism, which is needed to clarify shielding problems and to predict outage rates on overhead lines.

The phenomena leading to long air gap breakdown under negative polarity are more complex than the positive ones. Indeed, fast camera recordings in frame and streak modes coupled to current measurements allowed us to show that the negative discharge progresses non-continuously through regular and discrete 'steps' separated by time periods of about 10–20 μs [4–8]. Figures 4.1 and 4.2 sketch the main processes occurring during the development of a negative discharge in a long air gap. Between two consecutive steps, complex phenomena can be observed [6].

Generally, the phenomena preceding the breakdown in a long air gap under negative polarity involve two phases (figure 4.2): cold phases with positive streamers (D) and negative streamers (C) and hot phases with leaders. The latter are found in different forms:

- a negative leader propagating from the cathode toward the anode (A);
- a spatial leader (B) created within the gap and propagating towards the cathode (positive space leader); and
- a positive leader (F) developing from the anode towards the cathode.

Stems are a 'transient phase' between streamers and leaders.

doi:10.1088/978-0-7503-1236-3ch4

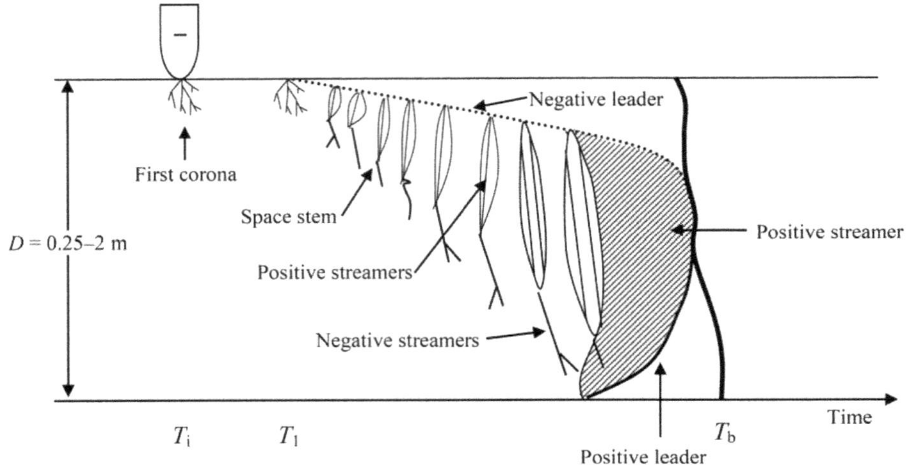

Figure 4.1. The sequence of the different phases of negative air gap breakdown; for distances lower than 2 m. Reproduced from [6] with permission of Université de Pau et des Pays de l'Adour, France.

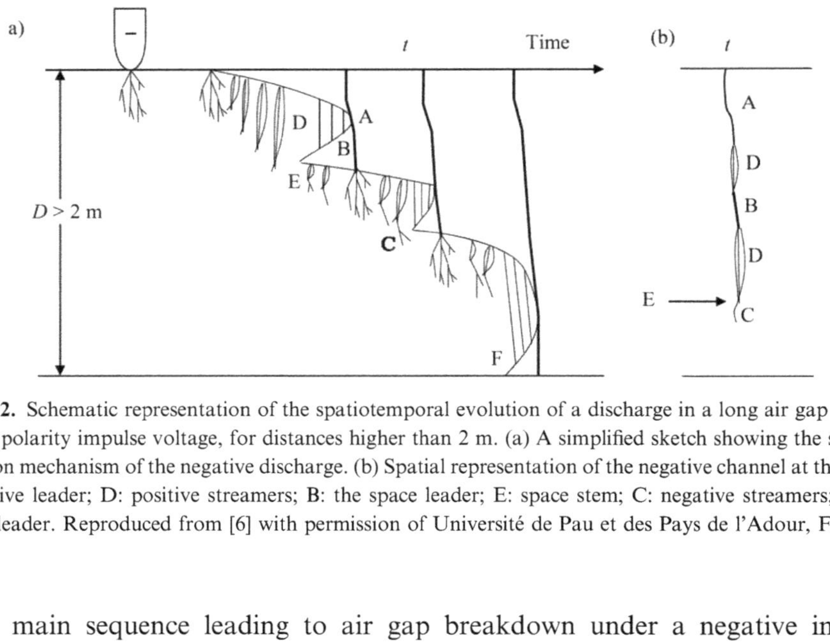

Figure 4.2. Schematic representation of the spatiotemporal evolution of a discharge in a long air gap under a negative polarity impulse voltage, for distances higher than 2 m. (a) A simplified sketch showing the stepping elongation mechanism of the negative discharge. (b) Spatial representation of the negative channel at the time t. A: negative leader; D: positive streamers; B: the space leader; E: space stem; C: negative streamers; and F: positive leader. Reproduced from [6] with permission of Université de Pau et des Pays de l'Adour, France.

The main sequence leading to air gap breakdown under a negative impulse voltage has five steps:

- first coronas (with or without dark periods) and stem inception;
- cathodic stem and upward discharge development;
- negative leader development;
- space stem or pilot system development; and
- the final jump and arc phase.

4.2 First negative corona and stem

The first phenomenon observed when the gap is subjected to a negative impulse voltage, is the first corona. This consists of a branched filamentary structure, similar to the first corona in positive polarity. Regardless of the polarity, the ionization process responsible for the dielectric breakdown of gases is given by the conditions described in section 3.2.1. However, under negative polarity, the statistical time lag is lower than that of the positive discharge and presents a much smaller dispersion (figure 4.3). The higher the rise time, the larger the dispersal is.

Following the inception of the first corona, the related charge flow causes a field reduction at the HV electrode and a concomitant increase in the field at the grounded-plane electrode.

Similarly to positive polarity, the first corona can be followed by one or many coronas with interspersed dark periods. Each corona is made up of a certain number of filaments called negative streamers which propagate from the cathode and develop into a cathodic stem. The corona filaments have a common root.

4.3 The cathodic stem and upward discharges

After the first coronas, an ionized zone in the gap, including a luminous nucleus called the *cathodic stem* by analogy with the *anodic stem* in the case of the positive polarity, is initiated. Its evolution into a negative leader is similar to that of the positive polarity. Its inception starts either after one or several dark periods, or coinciding with the first corona.

The development of positive streamers and stem formation are summarized in figure 4.4. The presence of positive space charges in negative streamers has been investigated by many authors (e.g. 3, 7). These positive ions were deposited in the wake of the developing negative streamers. These important positive space charges,

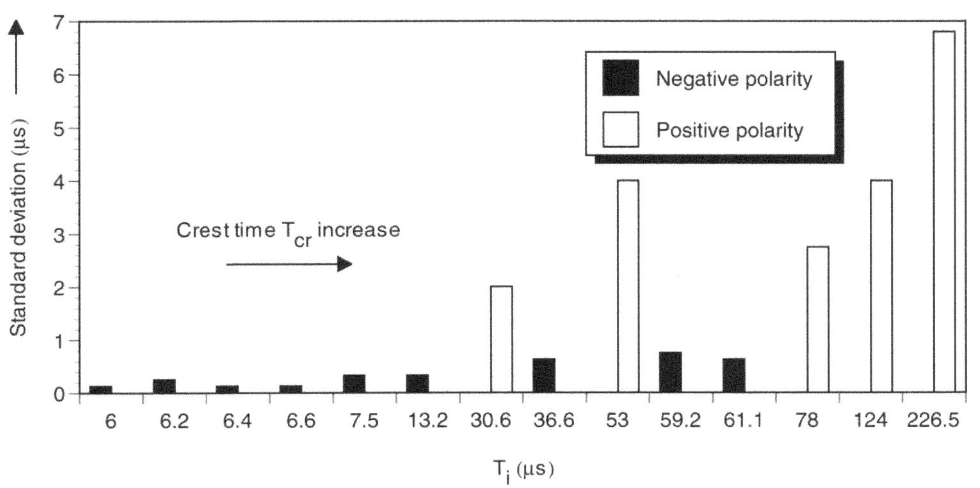

Figure 4.3. Variations of the standard deviation as a function of first corona inception time for both polarities, in the case of various wave shapes and an hemispherical HV electrode (ϕ = 600 mm). Reproduced from [6] with permission of Université de Pau et des Pays de l'Adour, France.

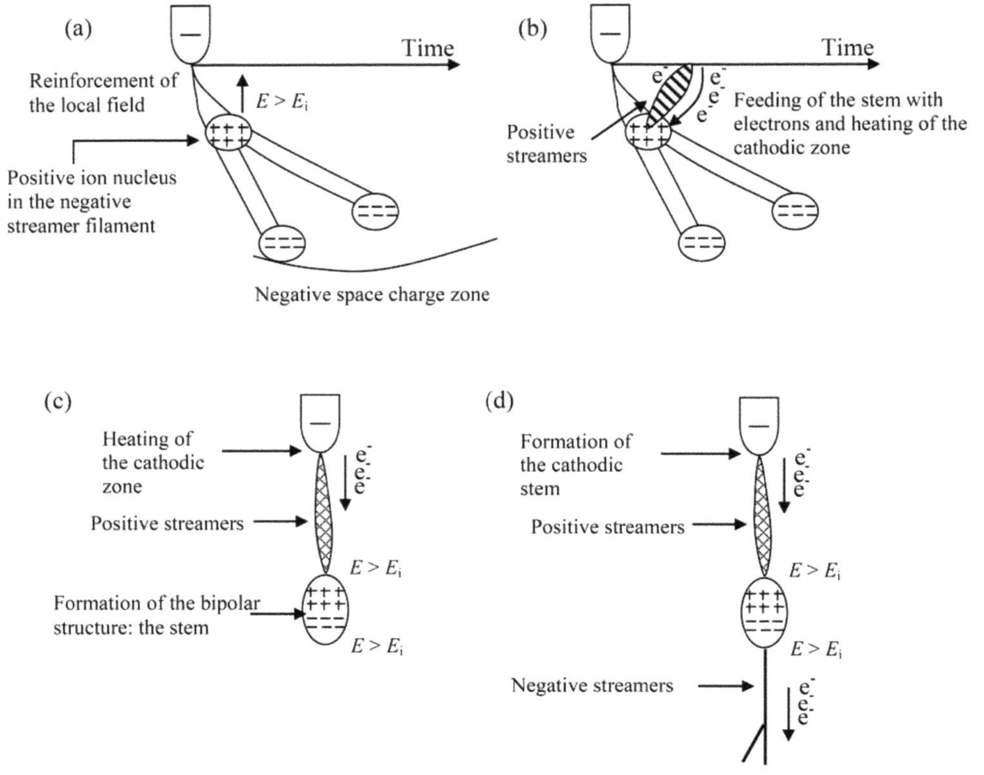

Figure 4.4. Schematization of the sequences leading to the formation of the pilot [6]: (a) positive space charge in the filament of the negative streamer; (b) positive streamers emission from the positive charge; (c) feeding of the stem in electrons and formation of the dipole; and (d) emission of a negative streamer and formation of the pilot.

which seem to be located at the root of negative streamers, initiate upward discharges that behave as 'positive streamers' drifting in a reverse direction towards the cathode (figures 4.4(a) and 4.4(b)).

The positive streamers generated from this positive space charge reach the HV electrode through the cathodic stem (the zone crossed by a significant number of electrons collected from the cathode) as illustrated in figure 4.4(c).

4.4 The negative leader

The electrons drifting through the positive streamers contribute to the heating of the cathodic stem. When the streamer emission is prolonged with time, the electron flux drifting through this cathodic stem is sufficiently high to transform this latter into a hot filament, called the negative leader. Similarly to the positive polarity, leader inception depends on the electrode geometry. With a small electrode radius, the leader inception time T_1 (figure 4.1) may not coincide with the corona inception time T_i (figure 4.1) similarly to the positive polarity.

The experimental investigations using special Schlieren techniques have shown that the leader properties are the same, regardless of the polarity [7]. This is why the majority of researchers believe that the physical models proposed for the positive polarity are also valid for the negative polarity.

The negative leader is mainly constituted by positive ions and electrons, where the reduced field E/N (E being the field along the channel and N the number of particles) and the electronic temperature are constants. This plasma is maintained by the flow of electrons that the cathode provides. LTE does not apply in the channel for the laboratory sparks. An erratic propagation of the channel is also depicted due to the space charge injected into the gap (see section 3.2.3.3). Ahead of the leader, an ionized zone called a 'space stem' leads to the discharge.

4.5 The space stem or pilot system

The *space stem* or *pilot system* which is the most advanced part in the gap has a more complex structure in comparison with the positive polarity. This space stem is formed in a region previously occupied by the corona-streamers. The space stem (or pilot system) (figure 4.4(d)) is made up of two coronas of opposite polarity, the positive one develops upwards to the HV electrode ('cathode-directed corona-streamers') while the negative one propagates downwards to the grounded plane ('anode-directed corona-streamers').

In the gap between the space stem and the anode, if electron emission from the positive streamers is high enough, i.e. if the stem is sufficiently fed with electrons, one or more negative streamers develop from the stem towards the grounded electrode. The feeding of stem with electrons is fundamental for negative streamer development (figure 4.5). It can be seen as a 'reserve of electrons' in the stem. It contributes to the electric field increase in the stem up to the onset level.

The length of a space stem is extremely variable. It may reach values up to 150 cm for air gap lengths of about 7 m [6, 7].

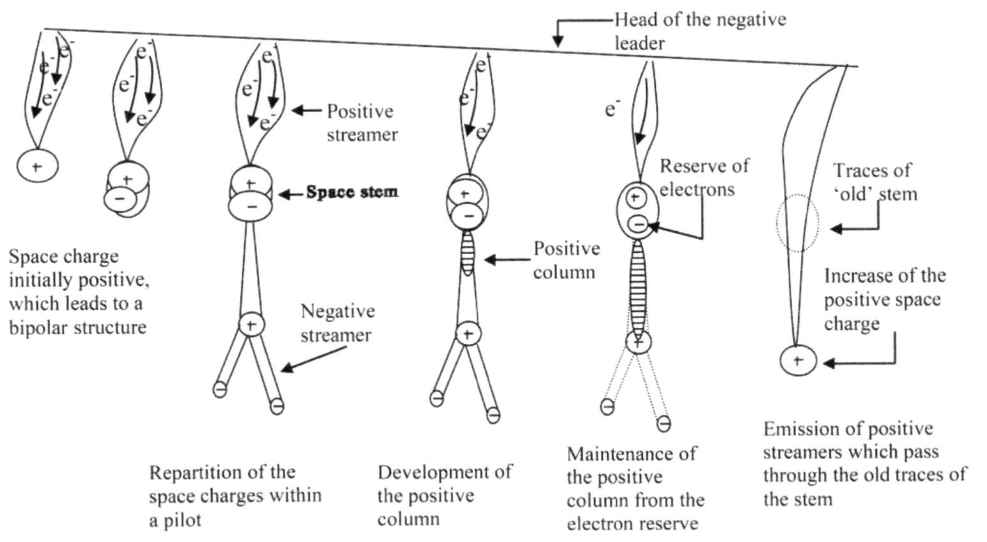

Figure 4.5. A sketch of the sequences leading to the mechanisms of replication of the space stem. Reproduced from [6] with permission of Université de Pau et des Pays de l'Adour, France.

As long as the space stem is able to sustain the positive column by feeding it electrons through the positive streamers, it will develop in the direction of the positive space charge abandoned by the negative streamers (figure 4.5). If the feeding in electrons is sustained, this positive column moves closed to the positive space charge which behaves as a floating anode.

When the electron reserve of the stem is no longer able to sustain the positive column, it cools down and stops progressing. The local field in the positive space charge increases enough to allow the emission of new positive streamers in the direction of the cathode. These positive streamers use the traces of the positive column and 'old' space stems which form an ionized medium, favourable for the streamer propagation.

The negative streamers appear as individual filaments, slightly branched. The positive streamers form a dense corona burst connecting the space stem with the negative leader. The propagation velocities of positive and negative streamers are of the same range in the vicinity of the space stem (a few cm μs^{-1}). Nevertheless, the velocity of the positive streamers decreases gradually as they move away from the stem.

The positive streamers propagate in the zone previously covered by the negative streamers, connecting the space stem with the negative leader tip with a dense corona burst. The negative streamers originating from a space stem play a determining role in the progression of the space stem towards the anode.

The whole system (positive and negative streamers) and the common space stem replicate in the gap like a 'bi-directional' corona from a virtual electrode. The equivalent propagation velocity towards the plane (grounded electrode) is about 10 cm μs^{-1} and depends on the steepness of the impulse voltage (dV/dt); it generally increases linearly with the steepness of the impulse voltage. Furthermore, the luminous activity of the space stem is more or less periodic, with a duration ranging from 0.3–0.5 μs with intervals of the same range.

4.5.1 Air gap $D < 2$ m

For an air gap distance of $D < 2$ m (as depicted in figure 4.6), the negative streamers reach the plane without initiation of a new discharge phase. The formation of the negative leader is instantaneously followed by the final jump phase (figure 4.1(a)). The sequence of this discharge looks similar to those observed under positive polarity, i.e. without inception of a space leader. Note that for waveforms for which the crest time is less than the critical crest time ($T_{cr} = 2$ μs for $D = 1$ m), although the stem spatial propagates by fits and starts, the discharge presents a quasi-continuous development [6]. For longer crest times, dark periods are observed in connection with which the spatial stem is reactivated.

4.5.2 Air gap $D > 2$ m

For much larger air gaps, $D > 2$ m (as depicted in figure 4.7), the space stem can develop relatively longer (before the final jump phase) and initiate a new phase in the discharge. Indeed, the streamers initiated from the space stem can develop into leaders.

The positive streamers develop rapidly in a leader-type channel (the space leader) which propagates from the spatial stem towards the cathode. The space leader

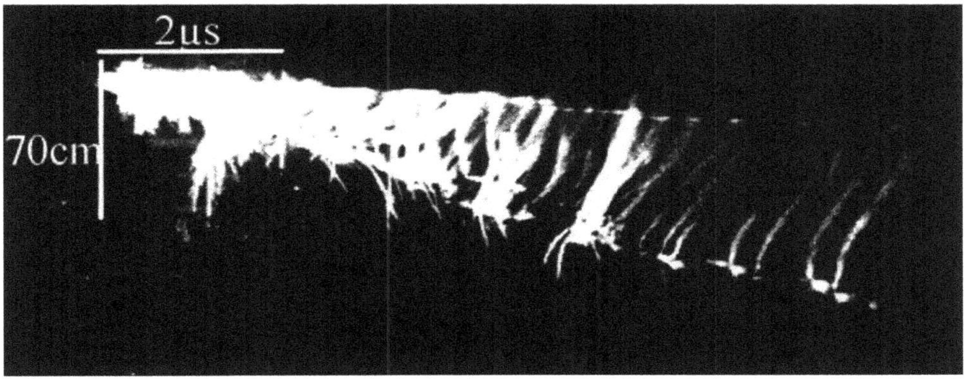

Figure 4.6. A streak photograph showing the inter-step phase of a negative discharge in a point–plane electrodes arrangement subjected to an impulse voltage of 6.5/1700 μs; $D = 1.3$ m; $R_p = 0.1$ cm; $U_{cr} = 1$ MV. Reproduced from [6] with permission of Université de Pau et des Pays de l'Adour, France.

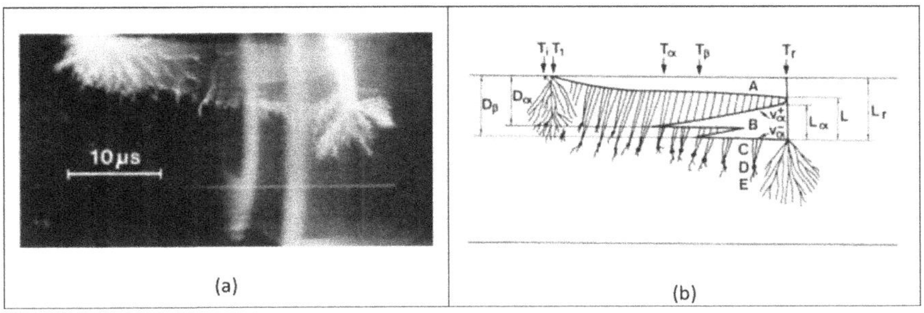

Figure 4.7. A streak photograph showing the inter-step phases of a negative discharge ($D = 7$ m, radius of curvature of point electrode = 30 cm, $V_{max} = 2.8$ MV; waveform 60/3000 μs) [7]. Figure reprinted with permission from CIGRE, Negative discharges in long air gaps at Les Renardieres Electra © 1981.

propagates as a bi-directional discharge. It is difficult to estimate the space-leader velocity due to the limited extension of the phenomenon in space and time, compared with the total discharge duration. Due to the limited accuracy in velocity determination, no special influence of the geometry has been observed experimentally [7]. The following axial velocities were estimated during stable propagation, (regardless of the geometry):

- v^- (anode-directed velocity) ranging from 0.7–1.5 cm μs^{-1} with an average value of 1 cm μs^{-1}.
- v^+ (cathode-directed velocity) ranging from 1–5 cm μs^{-1} with an average value of 3 cm μs^{-1}.

The distance between the space leader and the negative leader tip is bridged rapidly due to the higher velocity of the space leader, and the 'junction' of both leaders produces a strong illumination of the whole channel (starting from their junction point): the negative leader is lengthened by the space leader. A new corona starts at its tips to complete the step process. The entire phenomenon resumes and it

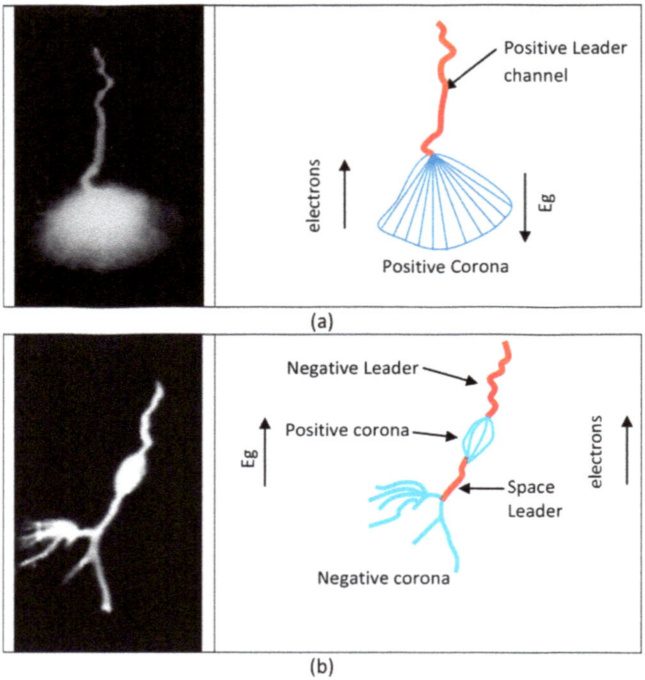

Figure 4.8. Positive and negative discharge structures with indication of the electron drift direction reproduced from [4].

is repeated until the discharge reaches the grounded plane. The inception of new corona causes an intensive burst of streamers propagating ahead, at least, up to the most advanced space stems, or beyond. Consequently some new space stems may be initiated. The space leaders which originated from these new space stems or an already existing one allow the repetition of this phenomenon. Since the intensity of the re-illumination process is highly variable, a junction of a space leader with a negative leader may occur without significant light emission.

The discharge process in both polarities may be considered quite similar [4]: a leader channel with a feeding discharge at its head. While in positive polarity the feeding discharge is simply a corona discharge, in negative polarity the structure is more complex, including pilot systems and space leaders (figure 4.8).

For the sake of analogy with positive polarity, the discharge phase involving the negative leader and one or more space leaders will be called the 'leader phase'. The average value of the axial velocity is approximately 1 cm μs^{-1}. No correlations with the shape and the inter-electrode gap have been observed [7].

4.6 Final jump phase

While in positive polarity the discharge channel mainly consists of a leader with a corona at its head, the structure appears to be more complex in negative polarity, as depicted in figure 4.9.

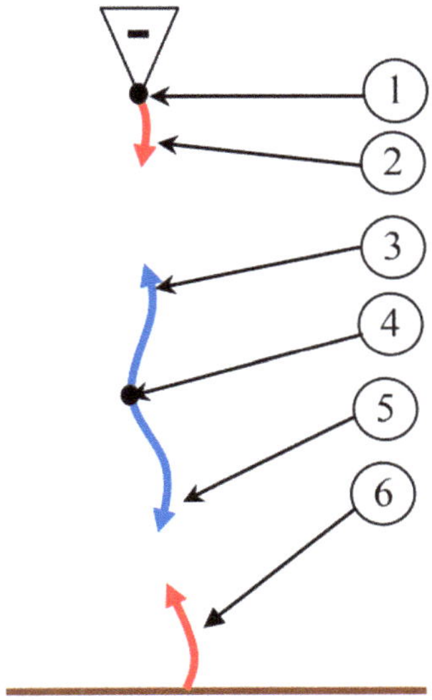

Figure 4.9. Schematic diagram showing the negative leader (2) issued from the stem (1), the space leader (3), (4), (5) elongating toward the cathode (3) and the anode (5), and the upward positive leader (6).

One may distinguish three leader channels before the final jump:

- a channel called the 'negative leader' developing from the cathode;
- a channel called the 'space leader' developing from a point in the gap towards both electrodes, which may cause an elongation of the negative leader with a restrike; and
- a channel called the 'positive leader' developing during the final jump, from a point of the plane (probably starting from an asperity or protrusion) towards the cathode.

For slow-front impulse voltages ($T_{cr} > 60$ μs), the discharge develops discontinuously before a negative streamer reaches the grounded plane; the emitted lights oscillate several times.

The final jump starts when the negative streamers reach the grounded plane. The arrival can occur either suddenly following leader elongation (re-illumination) or during the continued propagation of the leader and associated space stem. It is useful to distinguish three different mechanisms, A, B, and C, leading to the final jump (with reference to the negative leader).

In all cases, breakdown will always follow. At the beginning of the final jump phase, the positive leader starts from the plane. When it joins with the downward leader, the discharge channel becomes highly luminous and final breakdown of the gap occurs.

4.6.1 Type A final jump

In the type A final jump, the negative leader propagation experiences less re-illumination and the streamer zone, whose extension is more important, becomes predominant. In this case, the height of the final jump is quite important, about 70%–90% of the gap length [7]. The lowest values correspond to shocks that have a negative re-illumination leader before or early in the final jump.

The main channel does not include the space leader just before or during the jump final. Upon arrival of streamers at the grounded plane, light emitted by the entire channel increases and the space stem accelerates towards the plane. When the space stem reaches the plane, positive streamers develop upward, accompanied by a positive leader. We can then distinguish, from the anode to the cathode, an ascending leader, a bright area of positive streamers, and a negative leader (figure 4.10). Note that for smaller air gaps ($D < 2$ m), the final jump is generally of type A.

4.6.2 Type B final jump

In a type B final jump, the space leader is involved during the final jump. Similarly to the type A mechanism, the final jump starts at the arrival of the streamer at the grounded plane. Depending on the spatial position of the space leader, either near the plane or near the head of the negative leader, it may connect with the positive leader (a) or with the negative leader (b) as shown in figure 4.11.

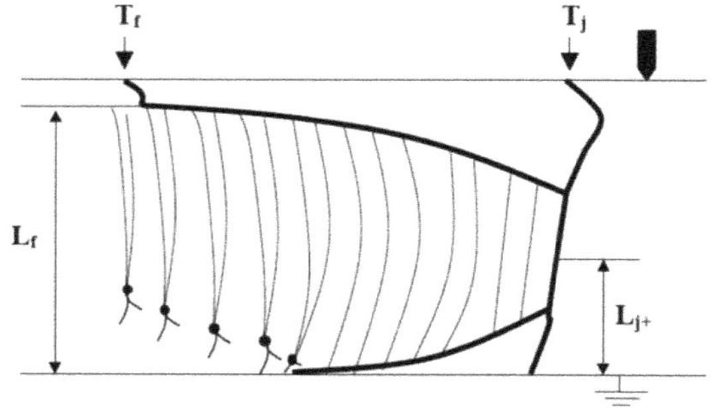

Figure 4.10. Schematic of the type A final jump reprinted from [7] with permission from CIGRE, Negative discharges in long air gaps at Les Renardieres Electra © 1981.

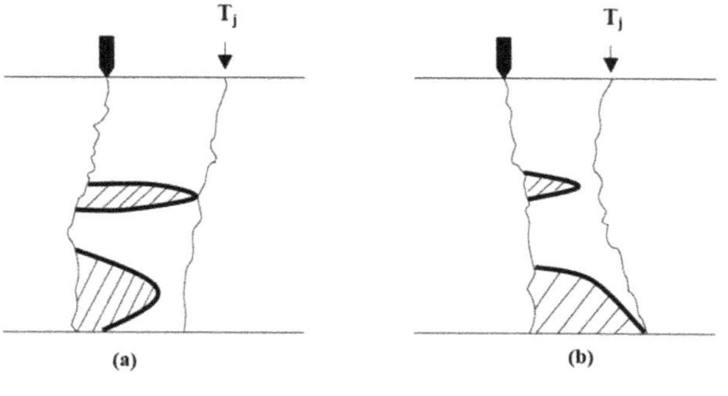

Figure 4.11. Schematic of the type B final jump: the space leader may join the positive leader (a) or the negative leader (b) reprinted from [7] with permission from CIGRE, Negative discharges in long air gaps at Les Renardieres Electra © 1981.

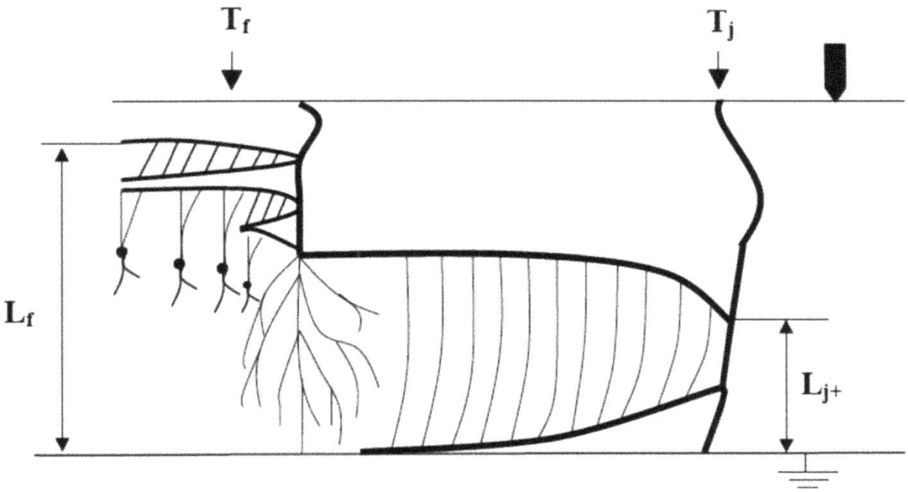

Figure 4.12. Schematic of the type C final jump reproduced from [7] with permission from CIGRE, Negative discharges in long air gaps at Les Renardieres Electra © 1981.

4.6.3 Type C final jump

The negative streamers can reach the plane after a sudden lengthening of the discharge channel at the joining of a negative leader and the space leader. This phenomenon is accompanied by a sudden elongation of streamers at the head of the leader and light emission. When the downward streamers reach the plane, upward streamers develop from the plane (from the impact zone). No space stem is visible during this phase. A positive leader is formed, similar to the type A mechanism (figure 4.12).

References

[1] Bazelyan E M and Raizer Y P 2000 *Lightning Physics and Lightning Protection* (Bristol: Institute of Physics Publishing)

[2] Lalande Ph 1996 Etude des conditions de foudroiement d'une structure au sol *PhD Thesis* University Paris XI

[3] Le Roy G, Gary C, Hutzler B, Lalot J and Dubanton C 1984 *Les Propriétés diélectriques de l'Air et les Très Hautes Tensions* (*Collection de la Direction des Études et Recherches d'Électricité de France* vol 51) (Paris: Eyrolles)

[4] Gallimberti I, Bacchiega G, Bondiou-Clergerie A and Lalande Ph 2002 *C R Phys.* **3** 1335

[5] Ortéga P 1992 Comportement diélectrique des grands intervalles d'air soumis à des ondes de tension de polarité positive ou négative *PhD Thesis* Université de Pau et des Pays de l'Adour, France

[6] Rees T 1997 Contribution expérimentale et théorique des décharges électriques de polarité négative dans l'air à pression atmosphérique *PhD Thesis* Université de Pau et des Pays de l'Adour France

[7] Les Renardières Group 1978 *Electra* **74** 67

[8] Les Renardières Group *Electra* **53** 31

Chapter 5

Lightning discharge

5.1 Introduction

Lightning is a phenomenon known to man for its spectacular and dangerous nature with bolt temperatures hotter than the surface of the Sun and shockwaves beaming out in all directions. This natural phenomenon occurs most frequently during thunderstorms, but has also been observed during volcanic eruptions, extremely intense forest fires, and surface nuclear detonations. The thunderbolt of mythology has long been feared as an atmospheric flash of supernatural origins: the great weapon of the gods [1, 2].

Benjamin Franklin (1706–90) is generally considered to be the 'father of modern lightning protection theory'. While other scientists had previously observed the similarity between laboratory sparks and lightning, Franklin was the first to provide the experimental evidence for the electrical nature of lightning. Prior to that, electrical science had developed to the point where positive and negative charge separation could be observed. Electrical instruments could, by rubbing together two different materials, store the charges in primitive capacitors (so-called Leyden jars), from which sparks could be generated and observed. T F D'Alibard of Marly-la-Ville in France successfully observed, in May 1752 (a few weeks before Franklin's experiment), sparks jumping from an iron rod during a thunderstorm [3]. G W Richmann, a Swedish physicist working in Russia in July 1753, proved that thunderclouds contain electrical charge and was killed when lightning struck him [2].

In Franklin's experiment, the kite took the place of an iron rod, since it could reach a greater elevation and could be flown anywhere. He theorized that clouds are electrically charged; consequently, lightning must also be electrical. During a Pennsylvania thunderstorm in 1752, the experiment involved Franklin standing on an electrical stand, holding an iron rod with one hand to obtain an electrical discharge between the other hand and the ground. The kite flew with sparks jumping from a key tied to the bottom of the damp kite string to an insulating silk ribbon tied to the knuckles of Franklin's hand. Franklin's grounded body provided a

doi:10.1088/978-0-7503-1236-3ch5

conducting path for the electrical currents responding to the strong electric field build-up in the storm clouds.

Several theories have been proposed to explain the separation or the build-up of electrical charges within clouds [2–8]:

- Simpson's theory: disintegration of water droplets;
- Findesein's theory: development, disintegration, and disappearance of ice crystals;
- Wilson's theory: accumulation of negative charges at water droplets falling under the Earth's electric field; etc

In fact, none of these theories is able to accurately explain charge separation in all cases. However, it is undeniable that electrical charge build-up results from strong upward drafts within clouds. Warm and wet air masses, lifted towards great heights, expand leading a drop in temperature and to an increase of the relative humidity of the air until drops appear and freeze. Generally, the processes of electric charge accumulation within clouds intensify with the heat and humidity content of the updraft. Such conditions appear during thunderstorms.

Since Franklin's experiment, significant progress has been made in improving our basic knowledge of lightning phenomena, assisted by the availability of photography and spectroscopics. Lightning research became particularly active in the late 1960s when the threat to power networks increased. Aerospace vehicles and solid state electronics (used in computers and other devices) as well as the improved measurement and observational capabilities (made possible by advancing technology) were some of the factors which helped intensify lightning research.

It is well established today that lightning is generated in electrically charged storm systems, but the process of cloud charging is still not fully understood. Each year, lightning is responsible for the death or injury of thousands of people and tens of thousands of animals in addition to a large number of electrical or phone outages and millions of dollars in property damage. Power networks are the structures most affected by lightning surges.

Despite the large number of works focussing on lightning over many decades, this phenomenon is still relatively poorly understood. This is largely due to the elusive nature of lightning, which may strike whenever and wherever.

In this chapter, the phenomenological aspects characterizing atmospheric discharges are described. The origin of lightning and its role in the maintenance of electrostatic equilibrium between the ground and the upper atmosphere is emphasized. A brief description of the different stages from a cloud–ground lightning discharge is also provided.

5.2 The global electric circuit

Earth can be thought of as a giant spherical capacitor, whose electrodes are constituted by the surface of the Earth itself and the electrosphere, the conductive zone of the atmosphere. In between these electrodes, the troposphere and the stratosphere act as the dielectric with a thickness of about fifty km. During fair weather, a potential difference of about 300 000 V exists between Earth's surface

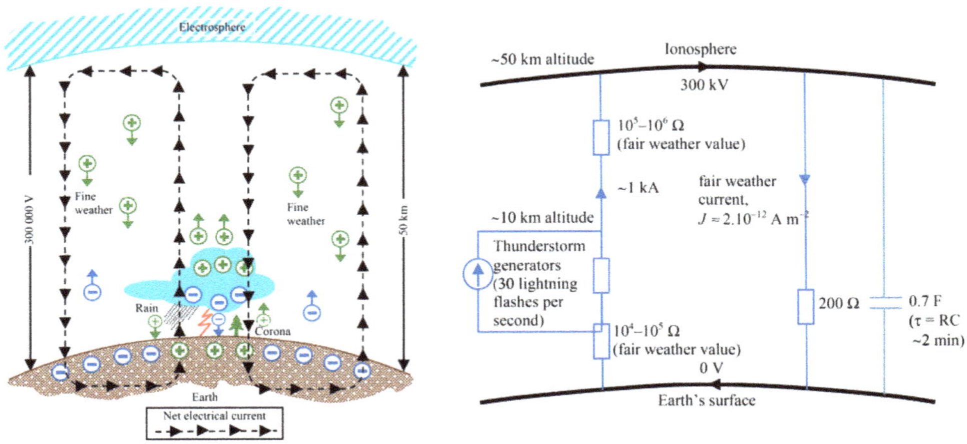

Figure 5.1. The spherical capacitance formed by Earth and the electrosphere reproduced from [9] with permission of Ecole Centrale de Lyon, France.

and the ionosphere. Experiments have shown that the electric field is greater in the middle of the day than in the morning or at night, and is also greater in winter than in summer. During fair weather, an electric field of 100 V m^{-1} exists at ground level and decreases with height. The electrosphere being positively charged, the electric field is directed toward Earth where the average negative charge distributed over the surface amounts to 10^6 C. A vertical current flowing through the electrically resistive atmosphere, shown in figure 5.1, was considered to be the source of the fair weather electric field by the Nobel Prize winner C T R Wilson in 1906. While current theory suggests that thunderstorms are responsible for ionosphere atmospheric current flow in fair weather, the details are still not fully understood. The atmospheric electrical circuit is modulated by variation in rain clouds and tropical thunderstorms. Modulation of the fair weather electric field occurs on daily, seasonal, solar cycle, and century timescales.

To maintain the fair weather global electric current flowing to Earth's surface at very low levels, 1000–2000 thunderstorms must be active at any time. Ground-based radio frequency measurements of global rates have significant uncertainties and limitations. A high resolution space based sensor is necessary in order to help improve our basic knowledge of the present uncertainties related to global lightning activity.

5.3 The most common types of lightning discharge

Lightning is a complex mixture of electrical and physical phenomena found in nature in different forms (figure 5.2). Typically, more than 2000 thunderstorms are active around the world at any time, covering about 10% of the surface of the Earth. Statistically, approximately 100 lightning strikes occur every second on Earth, adding up to nearly 8 million lightning flashes per day, 33% being a cloud–ground discharge. Basically, during a human life, the probability of being struck by lightning is one per 600 000. It should be emphasized that all types of lightning are dangerous and even the weakest thunderstorms produce lightning.

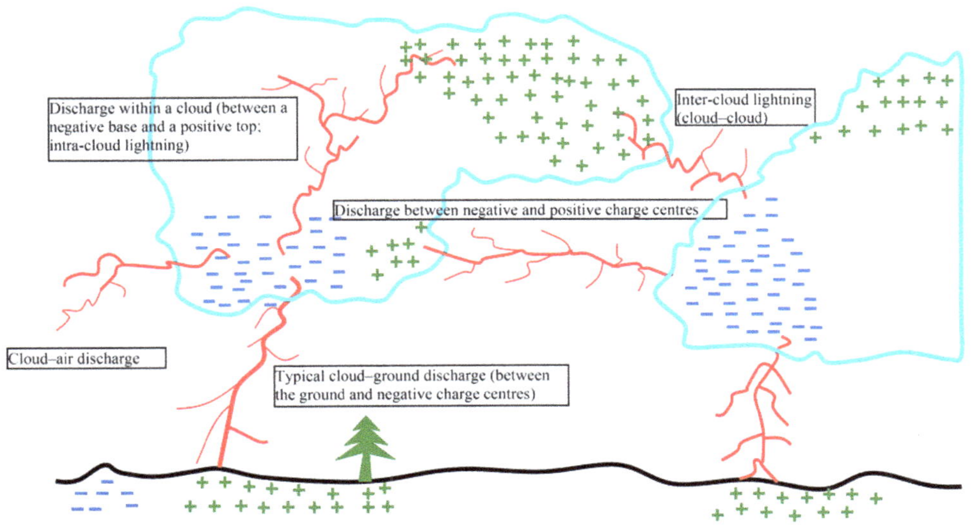

Figure 5.2. Schematic representation of typical atmospheric discharges. The lower part of a thundercloud is usually negatively charged in temperate regions while the upward area is usually positively charged.

On a small scale, all types of lightning are essentially the same–a channel of ionized air carrying electrical current between two differing areas of charge. The charge separation process takes place within a cloud. Thus far, the processes leading to cloud–ground or intra-cloud discharge have not been fully understood. However, researchers do believe that the discharge will occur within the cloud or progress to Earth depending on the height of the cloud above ground and changes in the electric field strength between the cloud and the Earth. If the field strength is highest in the lower regions of the cloud, a downward flash may occur from this region and propagate to the ground. The 'types' of lightning generally refer to how a discharge appears to an observer, what the starting and ending termini of the channel are, and in what direction the formation of the channel took place.

Cloud–ground lightning occurs between the cloud and ground. This is one of the most deadly natural phenomena known to man. Among the different types of lightning, this is the most studied and most understood. Figure 5.3 shows a typical cloud–ground lightning discharge. In temperate countries, most flashes (80%–90%) originate near the lower-negative charge centre and deliver negative charge to the ground. However, an appreciable portion of flashes (10%–20%) carry a positive charge to the ground. Positive flashes are also more common during winter days. In tropical zones, most cloud–ground flashes (80%–90%) carry a positive charge to the ground.

Cloud–ground lightning is classified in two groups according to the following criteria:

- The polarity of the ionized channel: lightning is positive or negative.
- The direction of the stepped leader: lightning is upward or downward.

It is therefore useful to define four types of lightning as depicted in figure 5.4, where v is the propagation velocity of the ionized channel and i is the lightning current. The configuration for temperate countries is depicted in table 5.1.

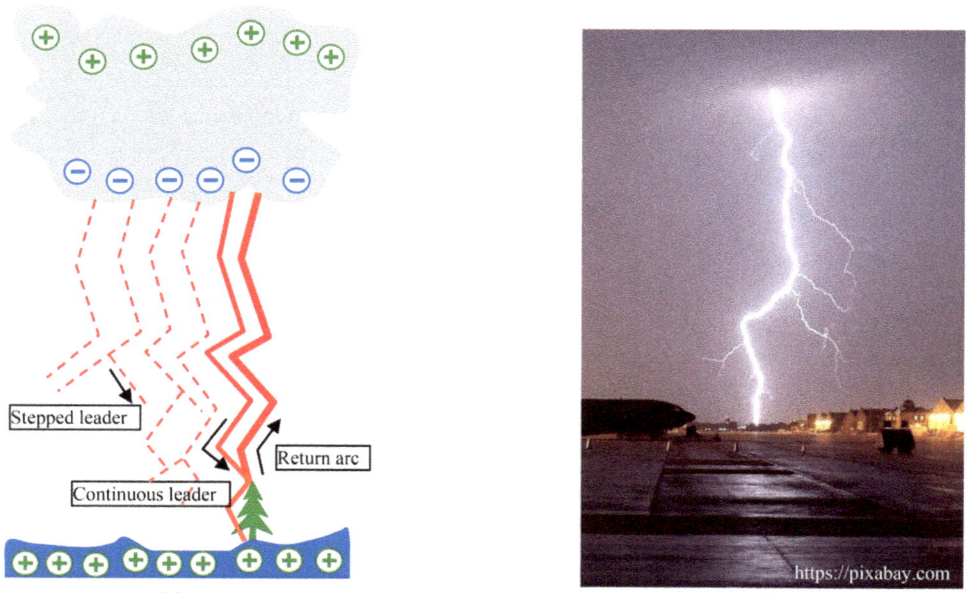

(a) (b)

Figure 5.3. Typical cloud–ground lightning discharge: (a) a sketch of a lightning strike on grounded structures and (b) a photograph of a typical cloud–ground lightning discharge. Courtesy of pixabay.com.

In recent years, most people killed by lightning were struck while swimming, golfing, or hiking. However, some have also been killed while doing less dangerous activities, such as talking on the phone, playing football or baseball, fishing on a lake, taking a shower, loading laundry into a clothes dryer, waiting for the bus, or walking.

Intra-cloud lightning is the most common type of flash (70%–80% of lightning flashes). This happens completely inside the cloud. Usually the process is seen from outside the cloud wherefrom it looks like a diffuse brightening channel which flickers. Intra-cloud lightning is sometimes called sheet lightning because it lights up the sky with a 'sheet' of light visible from a few kilometres away.

The ratio of cloud–ground to intra-cloud lightning can vary significantly between storms. Thunderstorms with a larger vertical height may almost exclusively produce intra-cloud lightning. Some researchers have suggested that changes are latitude-dependent, with a larger number of cloud–ground strikes occurring at higher latitudes. Other researchers suggest that the top height of the cloud is more important than latitude [4–10].

Cloud–cloud lightning is a less common lightning strike occurring between oppositely charged areas of two or more separate clouds.

Cloud–air lightning is an electrical discharge between a positively charged cloud top reaching out to the negatively charged air around it. This type of lightning is common in arid zones and during fair weather and is not as powerful as cloud–ground lightning.

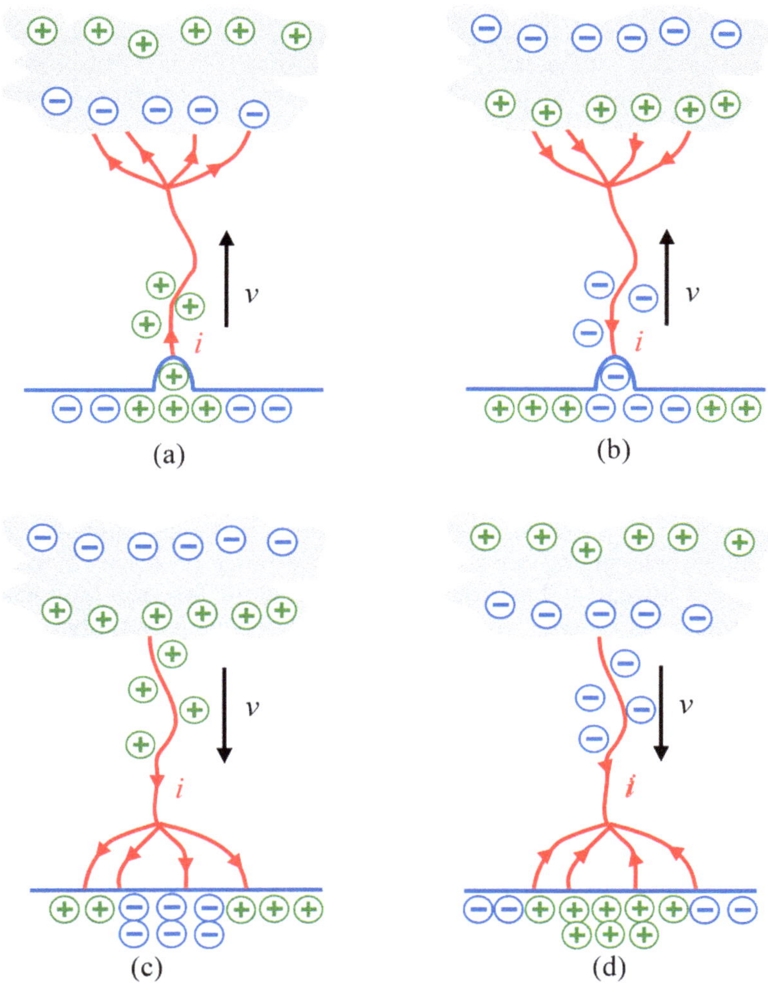

Figure 5.4. Cloud–ground lightning classification: (a) upward positive lightning (ionized channel +, lightning current −); (b) upward negative lightning (ionized channel −, lightning current +); (c) downward positive lightning (ionized channel +, lightning current +); and (d) downward negative lightning (ionized channel −, lightning current −).

5.3.1 Other types of lightning

Descriptions of various types and forms of lightning are provided in the literature [1–10]. Some identify subcategories and others may arise from optical illusions, appearances, or myths. Some of the most popular terms include the following.

Ball lightning is one of the unexplained atmospheric electrical phenomena. To date, many scientists still argue that ball lightning is not a real phenomenon but an urban myth. This type of lightning is described as a luminous ball of fire (of a few dozens of centimetres in diameter) travelling at high speed in the atmosphere. Very few scientific data on natural ball lightning are available owing to its scarcity and

Table 5.1. Characteristics of cloud–ground lightning discharges.

Types of cloud–ground lightning discharge	From tall objects	From clouds	Polarity of the leader	Conventional sign of lightning current	Figure
Upward positive lightning	✓		+	−	5.4(a)
Upward negative lightning	✓		−	+	5.4(b)
Downward positive lightning		✓	+	+	5.4(c)
Downward negative lightning		✓	−	−	5.4(d)

unpredictability. Its probable existence is based on reported public sightings and has therefore produced somewhat inconsistent findings. Given the inconsistencies and the lack of reliable data, the true nature of ball lightning is still unknown.

Sprites are massive but weak electrical discharges that occur high above thunderstorm cloud systems, giving rise to a quite varied range of shapes flickering in the night sky. They appear as luminous reddish-orange or greenish-blue flashes, with hanging tendrils below and arcing branches above their location, and can be preceded by a reddish halo. They are usually coincident with cloud–ground or intra-cloud lightning strikes at a 50–90 km altitude from the location of the lightning strike and with a time delay of up to 100 ms.

Blue jets are optical ejections from the top of the cumulonimbus above a thunderstorm. They are cones of dim blue light brighter than sprites and begin around an altitude of 15 km and rise to around 45 km like a quick puff of smoke. They are rare and different from sprites.

Elves, an acronym for emissions of light and very low frequency perturbations from electromagnetic pulse sources, refer to the process by which the light is generated; the excitation of nitrogen molecules due to electron collisions (the electrons possibly having been energized by the electromagnetic pulse caused by a discharge from the ionosphere). Though they can be accompanied by sprites, the causative mechanism is entirely different in nature. Elves often appear as dim, flattened, circular in the horizontal plane, expanding glows around 400 km in diameter that last for, typically, just one millisecond, which occur high above energetic cloud–ground lightning of positive or negative polarity. Some investigators have speculated that they may play a role in the formation of the ozone layer.

Heat lightning refers to a flash on the horizon or other clouds from distant thunderstorms that do not have accompanying sounds of thunder. This occurs because the lightning occurs very far away and the sound waves dissipate before they reach the observer. ‘Heat’ probably refers to the fact that this kind of display is often observed on hot humid summer nights, after daytime heating has triggered thunderstorms—just not near the observer.

Bead lightning, also called ‘chain lightning’, is a type of cloud–ground lightning which appears to break up into a string of luminous segments instead of a continuous channel, which last longer than the usual discharge channel. It occurs infrequently but has been observed many times. Its causes are unknown; several theories have been proposed to explain this phenomenon. Some authors theorized that bead lightning is a stage of a normal lightning discharge rather than a type of lightning in itself: the width of the lightning channel varies; parts of the channel are obscured by rain or clouds; wider sections cool more slowly and remain visible longer, appearing as a *string of beads* [5, 11].

Sheet lightning is an informal name for cloud–cloud lightning that occurs when the actual bolt or flash of lightning is hidden by clouds or terrain aside from the flash of light it produces. This form of lightning exhibits a diffuse brightening of the surface of a cloud, caused by the actual discharge path being hidden or too far away. Since the lightning itself cannot be seen by the observer, it appears as only a flash, or a sheet of light.

Ribbon lightning occurs in thunderstorms with high cross winds and multiple return strokes. This is typically caused by wind blowing the lightning channel sideways during exposure, resulting in a broadened appearance (ribbon effect).

Silent lightning occurs where airborne matter muffles the thunder, such as dust and sand storms and heavy snow in winter storms. In some instances, heavy falling snow has silenced thunder from cloud–ground lightning strikes as close as 1.6–3.2 km from the observer and severe dust storms are even more effective in many cases.

5.4 A description of a cloud–ground lightning discharge processes

Basically, lightning is formed as a result of the natural build-up of electrical charge separation in storm clouds. Cloud–ground lightning can be described schematically through the following steps [1–10, 12]:

- thunderstorm cloud (cumulonimbus) formation;
- charge separation in the cloud;
- preliminary discharge and leader propagation;
- return stroke and arc phase;
- dart (continuous) leader; and
- subsequent return strokes.

The ideal conditions for producing cloud–ground lightning and associated thunderstorms occur where warm, moist air rises and mixes with cold air at higher altitudes (2–7 km). These conditions occur almost daily on many parts of the Earth and rarely in other areas.

5.4.1 Storm cloud formation

Storm clouds, also known as cumulonimbus clouds, are cumulus clouds growing vertically instead of horizontally (figure 5.5). The base of a cumulonimbus can be several tens of square kilometres, and it can be tall enough to occupy the middle as well as low altitude. Its peak can reach up to 23 km in extreme cases with the lower end at an altitude varying between 2–7 km [5, 10, 13]. Cumulonimbus clouds are also characterized by a flat, anvil-like top (the anvil dome), caused by straight-line winds at higher altitudes. This gives the cumulonimbus its mushroom shape.

Figure 5.5. A cumulonimbus cloud. Courtesy of pixabay.com.

There are two types of storm clouds which generate a static electrical charge: heat storms and frontal storms. The thermodynamic phenomena involved in storm clouds are very complex and depend on many meteorological and physical factors: heat, humidity, and atmospheric turbulence.

Heat storms (figure 5.6) predominate in tropical regions and mountainous areas. On hot days, hot and humid air rising from the warm ground is replaced by cold air drifting down. The rising air, lifted by updrafts, condenses when it reaches greater altitudes (about 2 km) and forms clouds, first as water droplets and then at greater heights as ice crystals. The duration of this kind of cloud formation is very short, only 1–3 h.

The likelihood of this kind of storm mainly depends on ground surface conditions (topography): an easily heated surface, ground with a vertical relief, and conditions facilitating the moistening of warm air. The clouds of thus formed thunderstorms, under the influence of wind, are moved along the surface of the ground. From there, thermal clouds also move. However, their scale is limited. Their travel results from the ground topography and its characteristics.

Frontal storms (figure 5.7) predominate in temperate regions. This kind of storm is caused by an encounter between a front of cold air and a mass of hot and humid air which is lifted above the advancing cold front. As the hot and humid air rises the process described above is repeated but the resulting cumulonimbus clouds may, in this case, extend over several tens of kilometres in width with heights of between 7.5 km and 23 km [5, 10, 13]. The intensity of frontal thunderstorms depends on the local structure of the ground and topography. Unlike heat storm clouds, frontal storm cloud build-ups may persist for several days.

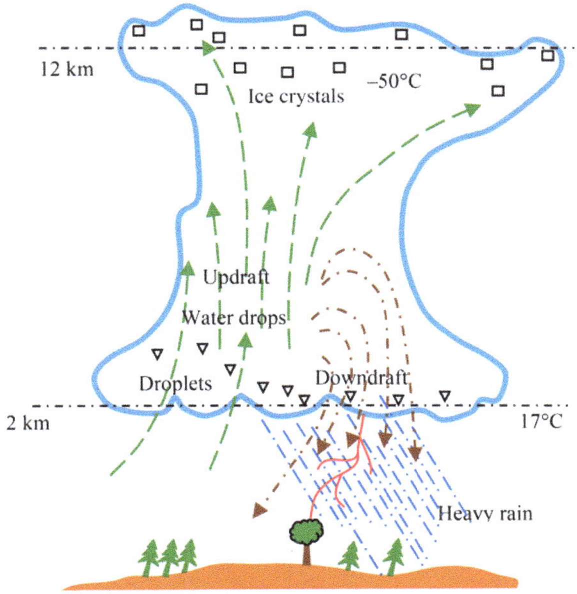

Figure 5.6. Heat storm formation.

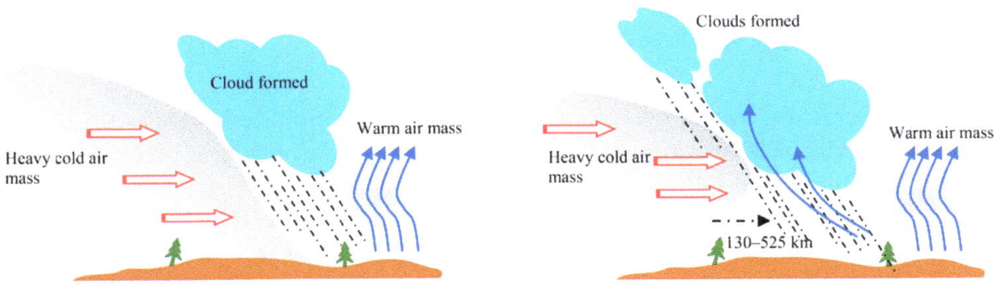

Figure 5.7. Frontal storm formation.

5.4.2 Charge separation within the cloud

How clouds form is well understood, but how the cloud separates its charge is not fully understood. The mechanism of charge separation within the storm cloud is certainly the least known step of all of the lightning processes, despite intensive research activity during the last century. Many different theories have been developed, but researchers do agree that in a thunder cloud, ice crystals become positively charged while water droplets carry a negative charge. This bipolarity is due to the ballo-electricity phenomenon which results in a charge transfer mechanism by friction between in-cloud particles (called hydrometeors).

These collisions are caused by atmospheric turbulence whose speed can exceed 20 m s^{-1} [8]. Under updrafts, the mechanism of water droplet condensation in the top of the cloud is reinforced. The low temperature at such an altitude freezes the droplets

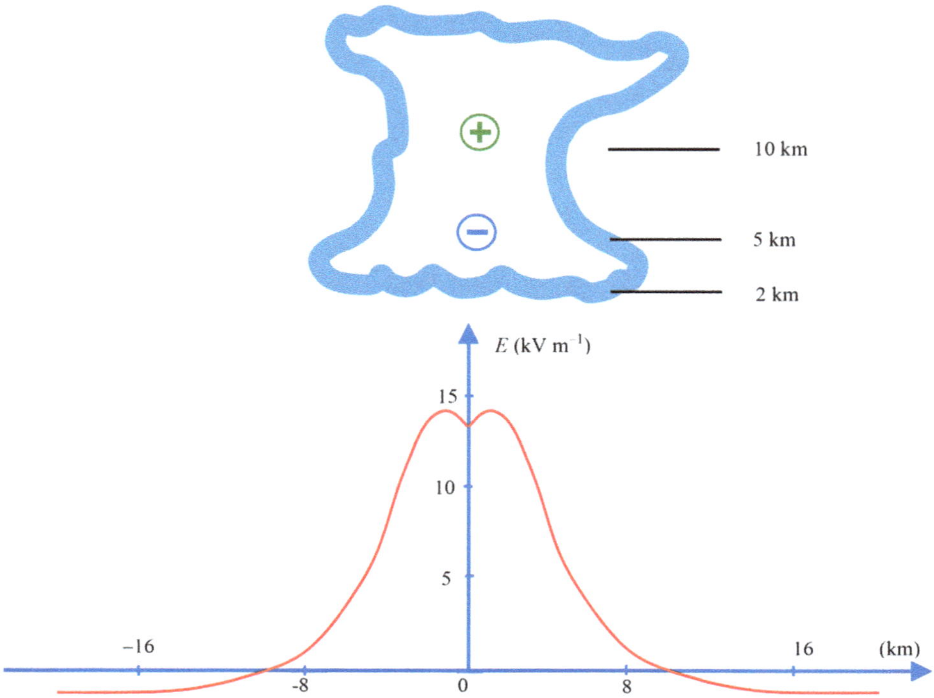

Figure 5.8. The electric field at ground level under a thunderstorm.

to form ice crystals. While these particles grow and precipitate by gravitation, light weight water droplets drift upwards to the top of the cloud. Friction between them is inevitable. During this friction, the water droplets lose their electrons which are caught by the frozen water droplets (Cohen's rule [8]). The charge separation by friction will continue under the influence of updrafts and gravity until the upper portion of the cloud acquires a net positive charge and the lower portion of the cloud becomes negatively charged. Under favourable conditions, the separation of charge produces a large voltage gradient, both within the cloud and between the cloud and the ground (figure 5.8) amounting to millions of volts.

At the end of this process, ice crystals (negatively charged) are deposited at the bottom of the cloud, while the water droplets (positively charged) eventually stabilize at the top. Often, a region of positive charges is inserted into the negative charges: the cloud is then a true dipole (figure 5.9). Consequently, the electric field at the ground increases. Once it reaches values between 15–20 kV m^{-1}, discharge to the ground is imminent. Assuming that this field is approximately uniform between the ground and the bottom area of the cloud, the voltage drop between both surfaces can reach a few tens of megavolts [14].

5.4.3 The lightning discharge sequence

The experimental study of lightning discharge development is possible only by means of high speed cameras. Figure 5.10 provides a simplified sketch of the

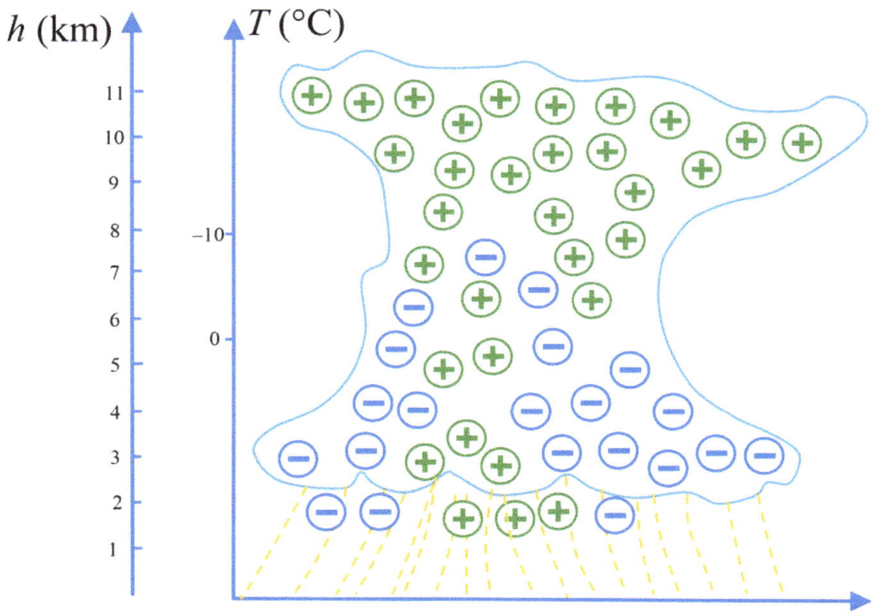

Figure 5.9. Schematic representation of a storm cloud and its electrical charge distribution.

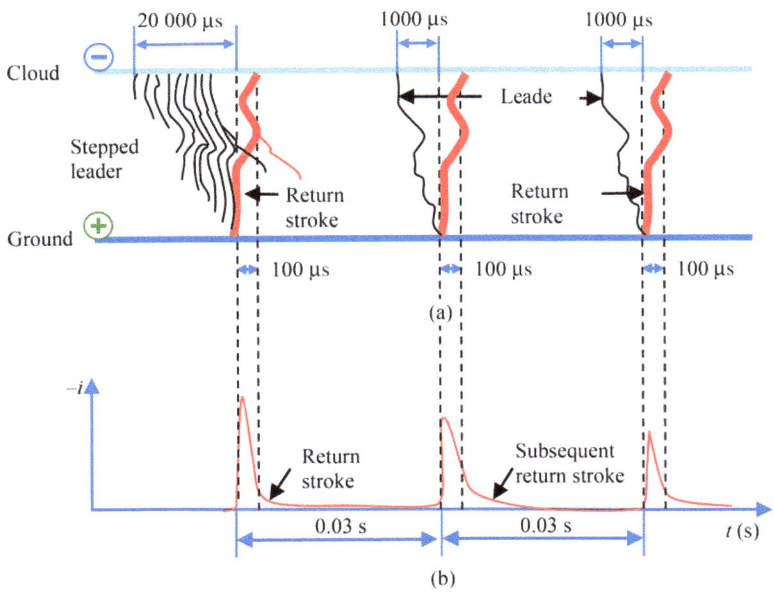

Figure 5.10. Lightning discharge development in the case of flat ground or a structure of moderate height at the ground: (a) a streak-camera-like sketch of the negative lightning discharge mechanism and (b) time variation of negative lightning discharge.

development of a cloud–ground lightning discharge. The considered case corresponds to a downward discharge between the cloud and a flat or moderate height (lower than 100 m) ground surface. The lightning discharge begins with a preliminary discharge within the cloud, in the storm centre.

5.4.4 Preliminary discharge and leader propagation

Several spectroscopic recordings have revealed the existence of several discharges in the lower part of the cloud at about 0.1 s, just before the inception of the first arc (stepped leader). These discharges are known as the 'preliminary discharges'. The generation of these discharges is possible only when the critical field of air is reached [8]. Most of our knowledge about the formation and development of this leader is based on the works by Schonland [6] and Krider [15].

With the initial breakdown of the air in a region of strong electric fields, a preliminary discharge may start and propagate downward toward Earth (figure 5.10). This discharge, called the stepped leader, propagates in discrete steps of about 3–50 m with an average velocity of about 150–200 km s^{-1}. The duration between two steps is about 40–100 μs. The leader describes an ionized path along which space charges are generated. As the stepped leader approaches the ground, a large potential difference is generated between the tip of the leader and the ground. The exact impact point is defined when approaching the ground (20–200 m). This distance, called the 'striking distance', is a function of the Earth's surface geometry and the total charge carried by the downward leader.

5.4.5 The return stroke

Typically, a streamer is launched from the ground structure that intercepts the downward stepped leader just before it reaches the ground. In the case of a thunderstorm that has grown over a tall grounded structure, such as a radio antenna, an upward leader may propagate from this structure toward the cloud. Once a connecting path is achieved, a return stroke flows up in the ionized path at speed roughly equal to $0.1c_o$ (c_o being the speed of light). Return stroke currents recorded during natural flashes range from 10–200 kA [4–10, 12]. The total charge drained to the ground varies between 5 and 15 C [4–10, 12]. The current flowing through the lightning channel induces a circular magnetic induction B (figure 5.11).

Making the assumption that the column is a perfect cylindrical one with a current density J, the magnetic self-striction which is applied on it can be expressed as a function of the pressure gradient, taking into account the magneto-hydrodynamic level of approximation [16]:

$$\nabla\vec{p} = \vec{J} \times \vec{B}. \tag{5.1}$$

The plasma is considered as a unique fluid on which a compressing force is applied. The order of magnitude of this force is given by Maecker *et al* [17]:

$$\nabla p = \frac{\mu_0 I^2}{4\pi^2 r_a^2}\left[1 - \left(\frac{r}{r_a}\right)^2\right] = p_M\left[1 - \left(\frac{r}{r_a}\right)^2\right], \tag{5.2}$$

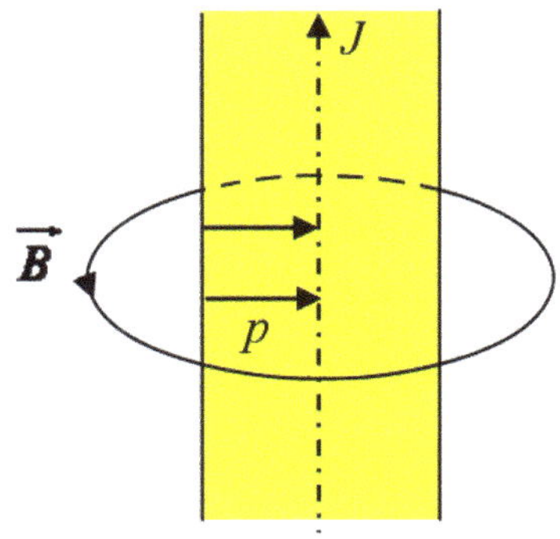

Figure 5.11. The magnetic force exerted on the cylindrical column of current density J, assumed to be uniform.

where p_M is the maximum pressure, I is the current flowing through the discharge, r_a is the conducting radius, and r is the distance from the discharge axis. This force tends to confine the column.

While lightning current is flowing in the channel, electro-dynamic forces produce channel contraction. The increase of the pressure inside the channel can reach 2–8 atm (10^5 Pa). This return stroke releases a tremendous amount of energy, bright light, and thunder. Sound is generated along the length of the lightning channel as the atmosphere is heated by the electrical discharge to the order of 30 000°C (four to five times the temperature of the surface of the Sun!). This compresses the surrounding clear air producing a shock wave, which then decays to an acoustic wave as it propagates away from the lightning channel. A lightning flash is composed of a series of four strokes. The length and duration of each lightning stroke varies with a typical average of about 30 μs.

5.4.6 The dart leader

When the current of the first discharge has stopped flowing, it may take about 100 ms [5, 8, 13, 18] before a second discharge occurs; the cloud may still contain a certain amount of charge: a condition required for a new flash. This second discharge starts with the propagation of a leader without branching, through the main channel of the previous leader, at a speed of about 10^7 m s^{-1} [5, 10, 14, 19].

5.4.7 The subsequent return stroke

As the dart (continuous) leader approaches the ground, opposite charges are pulled to the ground and then drained toward the base of the tracer in the form of a new upward discharge called the subsequent return stroke (figure 5.10). This discharge is characterized by a continuous un-branched path (the channel is already in place) and a propagation speed estimated between 1 and 1.2 10^8 m s^{-1} [5, 14, 18–20].

5.5 Lightning electrical parameters

The electrical parameters of lightning (potentials and currents) depend on the location. Using satellites orbiting the Earth with sensors designed to detect lightning

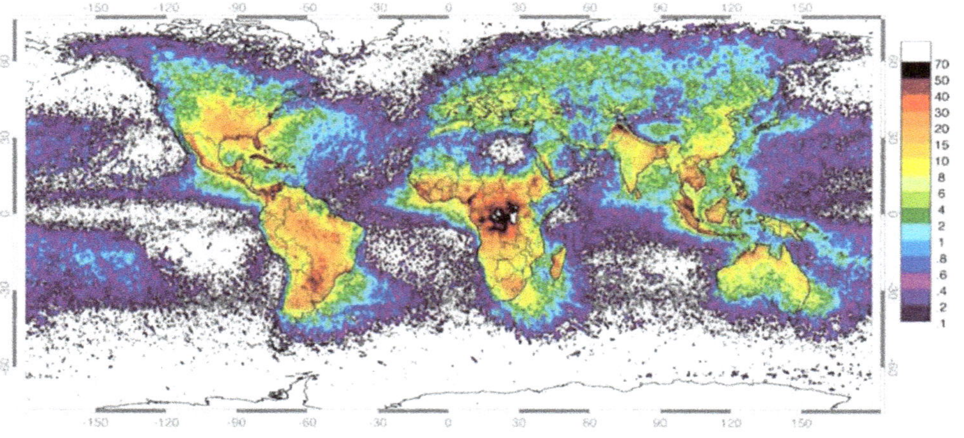

Figure 5.12. Global distribution of lightning April 1995–February 2003 from the combined observations of the NASA OTD (4/1995–3/2000) and LIS (1/1998–2/2003) instruments. The map is based upon the average yearly count of lightning flashes per square kilometre reproduced from [21] courtesy of NASA.

and different recording and visualization systems (optical transient detectors (OTDs), lightning imaging sensors (LISs), etc) it has been possible to map the annual lightning flash rate (figure 5.12). The distribution of lightning on Earth is far from uniform.

The lightning current polarity may be positive or negative. The average bolt of negative lightning carries an electrical current of about 30 kA and transfers 15 C of electrical charge and 500 MJ of energy. Large bolts of lightning can carry up to 120 kA and 350 C [22]. Positive lightning flashes may carry an electrical current up to 300 kA—about ten times the average value [23].

The shapes of the currents associated with negative lightning strikes vary both in amplitude and duration. However, for the first return stroke, the rise time of the current pulse is of the order of 5–10 μs and the duration at mid-amplitude is about 100 μs [13]. The magnitude of the peak current value is several tens of kiloamperes. The rise time of the subsequent return stroke current is less than a microsecond, with a lower amplitude compared to the first return stroke.

Positive lightning strikes are generally composed of a single discharge (without subsequent strokes) with a total duration ranging from 0.1–0.2 s. The rise time of the current pulse is relatively long (20–50 μs) while the current amplitude can reach values higher than 100 kA [9]. For example, figure 5.13 depicts some positive and negative lightning currents measured at Mt San Salvatore in Lugano [24].

In addition to the current wave forms, two other important parameters need to be emphasized. One of them is the total charge Q neutralized during a lightning strike. The average value is of the order of tens of coulombs, but in some cases it may reach values up to 300 C [9]. The second parameter refers to the specific energy, related to $\int i^2 \mathrm{d}t$. It is a parameter required for the evaluation of the thermal effects of lightning or

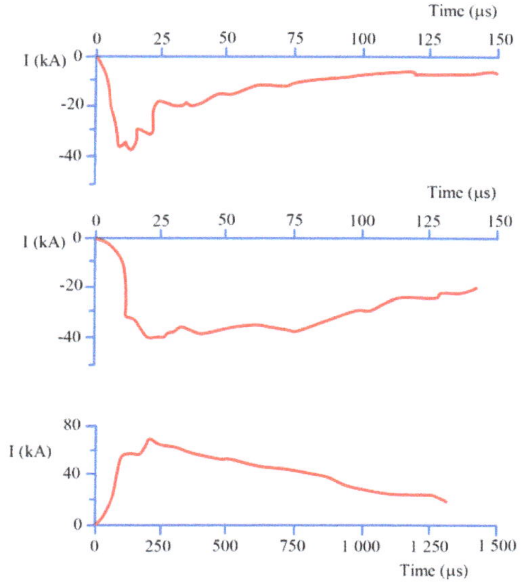

Figure 5.13. Positive and negative lightning current measured at Mt San Salvatore in Lugano (Switzerland) reprinted with permission from [24].

lightning rod conductor sizing. Changes in current ($i(t)$, $\mathrm{d}i/\mathrm{d}t$) or voltage ($V(t)$, $\mathrm{d}V/\mathrm{d}t$) are also important for electromagnetic compatibility or electrostatic discharge studies.

Table 5.2 summarizes the electrical characteristics of lightning (positive and negative polarity).

Figure 5.14 provides the experimental statistical distribution, in amplitude, of positive and negative lightning strokes [26].

5.6 Comparison of laboratory sparks and cloud–ground lightning discharges

Laboratory sparks can be compared only with the first component (discharge) of the lightning strike. From a purely phenomenological point of view, lightning and large sparks both result from the propagation of discharges of identical structure: a channel called a leader ending in a corona made of streamers.

Two categories of differences can be observed. The first difference concerns the order of magnitude of the various parameters. For instance, the current flowing in the lightning leader reaches 100 A, with a linear charge density estimated at 800 μC m^{-1} compared with a few tens of μC m^{-1} for long laboratory sparks. Table 5.3 shows some example electrical parameters for the two kinds of discharge.

However, researchers think that increasing the air gap length might bring air gap discharge closer to the magnitude of lightning characteristics, without revealing new phenomena. It is therefore of interest for physicists to base their fundamental research and models on experimental results obtained with laboratory sparks which are more convenient to observe.

Table 5.2. Statistical electrical parameters of lightning flashes [25].

Parameter	Unit	Probability		
Amplitude		95%	50%	5%
First stroke	kA	7	33	85
Subsequent stroke		4.6	12	30
Positive lightning		4.6	35	250
Max. slope	kA μs^{-1}			
First stroke		9.1	24	65
Subsequent stroke		10	40	162
Positive lightning		0.2	2.4	32
Specific energy	kA^2 s			
First stroke		6.0	55	550
Subsequent stroke		0.55	6.0	52
Positive lightning		25	650	1.5×10^4
Total charge	C			
Negative lightning		1.3	7.5	40
Positive lightning		20	80	350
Total duration	ms			
Negative lightning		0.15	13	1100
Positive lightning		14	85	500
Number of re-illuminations	n	1	2	6

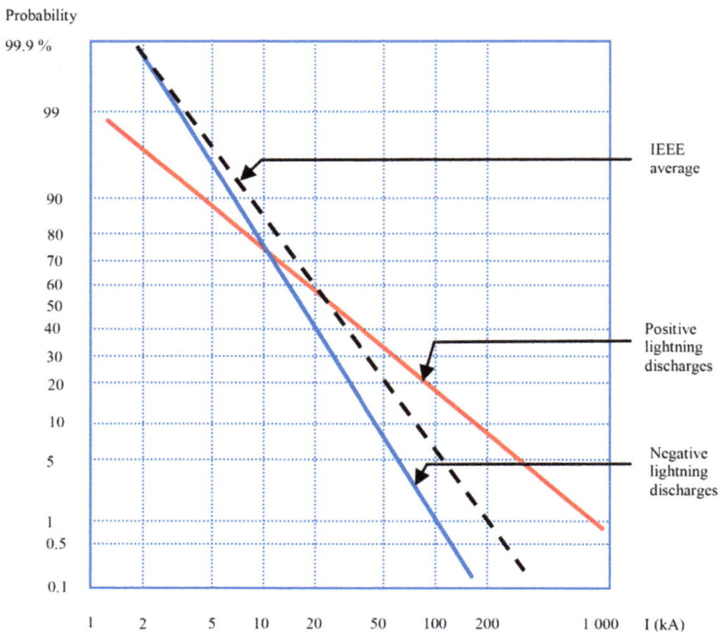

Figure 5.14. Statistical distribution, in amplitude, of positive and negative lightning strokes reprinted with permission from [26].

Table 5.3. A comparison of lightning and laboratory sparks [26].

		Lightning	Laboratory spark
Positive leader	Velocity (cm μs^{-1})	10–100	1.5–2
	Current (A)	50–200	0.5–1
Negative leader	Mean velocity (cm μs^{-1})	10–50	5–10
	Mean current (A)	–	5–10
	Duration between steps	30–50	10–20
	Length of steps (m)	5–20	0.5–1.5

The second difference is related to the qualitative aspect, since in the laboratory, one can observe complex phenomena between steps (the space stem and space leader), phenomena not highlighted in the case of lightning leaders. This difference can be explained by the conditions of observation which are much easier in the laboratory. An apparent difference does not necessarily mean a different *physical nature*. Although the absolute identity of the physical nature remains to be demonstrated, there is strong evidence that within a few-metre-long air gap breakdown, the structure of the discharge is the same (leader–streamer) as for lightning. From this point of view, models developed for large sparks could be extrapolated to atmospheric discharges.

References

[1] Bouquegneau C 2006 *Doit-on Craindre la Foudre?* (Les Ulis: EDP Sciences)
[2] Gary C 1994 *La Foudre: des Mythologies Antiques à la Recherche Moderne* 2nd edn (Paris: Masson) p 436
[3] Krider E P 2004 Benjamin Franklin and the first lightning conductors *Proc. Int. Commission on History of Meteorology* vol 1.1 (Mexico City: ICHM) p 1–13
[4] Le Roy G, Gary C, Hutzler B, Lalot J and Dubanton C 1984 *Les Propriétés Diélectriques de l'Air et les Très Hautes Tensions* (*Collection de la Direction des Études et Recherches d'Électricité de France* vol 51) (Paris: Eyrolles)
[5] Uman M A 2001 *The Lightning Discharge* (*International Geophysics Series*) (New York: Dover)
[6] Schonland F J 1956 *The Lightning Discharge* (*Handbook der Physik* vol 22) (Berlin: Springer)
[7] Betz H D, Schumann U and Laroche P 2009 *Lightning: Principles, Instruments and Applications—Review of Modern Lightning Research* (Berlin: Springer)
[8] Aguet M and Ianoz M 1987 *Haute Tension* (Paris: Dunod)
[9] Rakotonandrasana J H 2008 Modélisation de la longue décharge négative – Application à la foudre *PhD Thesis* Ecole Centrale de Lyon, University of Lyon
[10] Ben Rhouma A 1996 Caractérisation et modélisation des champs électromagnétiques rayonnés par les décharges de foudre en vue de l'élaboration d'un simulateur expérimental *PhD Thesis* Ecole Centrale de Lyon, University of Lyon
[11] Glickman T S 2000 *Glossary of Meteorology* 2nd edn (Boston, MA: American Meteorological Society)
[12] Bazelyan E M and Raizer Y P 2000 *Lightning Physics and Lightning Protection* (Bristol: Institute of Physics Publishing)

[13] Uman M A 1971 *Proc. IEEE* **59** 457

[14] Fofana I 1996 Modélisation de la décharge positive dans les grands intervalles d'air *PhD Thesis* Ecole Centrale de Lyon, University of Lyon

[15] Krider E P, Wiedman C D and Noggle R C 1977 *J. Geophys. Res.* **82** 951

[16] Hoyaux M F 1968 *Arc Physics* (New York: Springer)

[17] Maecker H and Popp H-P 2009 *The Electric Arc: the Physics of Stationary Gas Discharges Near Thermal Equilibrium* (Berg: Popp)

[18] Hutzler B 1988 *Bull. Dir. Etud. Rech.* **3** 31

[19] Gary C, Cimador A and Fieux R 1975 *Rev. Gen. Electr.* **84** 24

[20] Darchérif A 1990 Contribution à la modélisation des phénomènes d'interfaces électromagnétiques dans les réseaux électriques *PhD Thesis* University of Grenoble

[21] Klobassa B 2012 Lightning protection and grounding networks solutions for wireless networks *BICSI Fall Conf. and Exhibition* (*Anaheim, CA, USA, September* 16–20)

[22] Hasbrouck R 1996 *Sci. Technol. Rev.* **5** 7

[23] Fulchiron D 1992 Surtensions et coordination de l'isolement *Cahier Technique Merlin Gérin ERI Lyon* **151**

[24] Berger K, Anderson R B and Kroninger H 1995 *Electra* **41** 23

[25] Uman M A 1987 *The Lightning Discharge* (*International Geophysics Series* vol 39) ed W L Donn (San Diego, CA: Academic)

[26] Berger K, Anderson R B and Kroninger H 1975 Parameters of lightning flashes *Electra* **41** 23

IOP Publishing

Discharge in Long Air Gaps
Modelling and applications
A Beroual and I Fofana

Chapter 6

A review of existing mathematical models developed for long air gap discharges

6.1 Introduction

Numerical modelling and simulation play an important role in almost all fields of engineering. The development of mathematical models both to assist the physical understanding of discharge in long air gaps and to predict data useful for electrical engineers has a history almost as long as the study of the breakdown phenomenon itself. From the engineering point of view, the ultimate goal of simulation/modelling is to predict the behaviour of large air gaps from purely macroscopic data. This task should begin with macroscopic data analysis that includes, in this case, voltage characteristics, the gap configuration, atmospheric conditions, criteria for the start of propagation, etc. Self-consistent mathematical models that combine rigorous mathematical treatment of all the mechanisms involved are therefore necessary. In the case of discharges in large air gaps, this level of simulation is not achieved because of the complexity of the phenomena and their mathematical formulation, and numerical difficulties in their solution. Modelling may allow the coupling of numerical simulations with experimental tests in order to reduce the number of tests—usually very expensive—required to test and validate new configurations. This important step (simulation + experiment) will be developed during the design of new configurations; numerical modelling may be used as a design tool. The development of such an ambitious tool that will allow the closing of some gaps in our basic understanding of long air gap discharges is still a challenge.

A few mathematical expressions have been proposed to predict some macroscopic discharge parameters [1–10]. Some of these are purely empirical and valid only under certain conditions which must be accurately defined. One must be careful in handling an empirical model because of the risks of straying outside its field of validity. Most of them are limited to positive polarity and a single phase, and can basically not be combined to form complete models. Apart from the models

doi:10.1088/978-0-7503-1236-3ch6

developed by the authors (chapters 7 and 8), few models that allow outlining of the behaviour of an air gap under specific conditions have been proposed [6, 7, 11–13].

Despite the progress in the last five decades in this field, several issues still need to be explored in order to inter-relate the multiple factors involved in the mechanisms of the initiation and development of electrical discharges in large air gaps. In this section, the main mathematical models that have been suggested over the years are reviewed.

6.2 Positive discharge models

Laboratory investigations performed on long air gaps allow the collection of new information on the mechanisms involved in discharge phenomena. Various models are still being suggested by researchers. These models aim to describe a particular phase of the discharge (corona, formation or development of the leader, final jump, etc), or the total discharge development with the succession of different phases. Since long air gap discharges appear as a sequence of phases with specific electrical characteristics, such models are reviewed in the following.

6.2.1 Corona related models

Critical size of the avalanche

Different models have been proposed to define the critical size of the avalanche leading to the triggering of a corona-like discharge. When the avalanche develops according to the streamer mechanism, as in the case of a HV rod–plane under positive polarity, the critical size of the avalanche can be defined by one of the following criteria:

- *Meek criterion*: the avalanche develops into a streamer when the field at its head increases to the level of the applied field [4].
- *Raether criterion*: the number of electrons produced by the primary avalanche must be at least equal to 10^8 [14].
- *Gallimberti's criterion*: the existence of a minimal number of positive ions within the head of the avalanche is a condition for further development into a streamer. The number of positive charges N^+, required for the stable development of the streamers, depends on the geometrical electric field E_g at the surface of the HV electrode; it is given by the following relationships [6, 15]:

$$N^+ = 0.558 \times 10^8 - 0.231 \times 10^3 E_g \text{ for } E_g \leqslant 20\text{kVcm}^{-1}$$

and

$$N^+ = 3.34 \times 10^8 \exp[-1.614 \times 10^{-5} E_g] \text{ for } E_g \geqslant 20\text{kVcm}^{-1}.$$

First corona inception

The first corona can be modeled in three distinct aspects:

- A study of the minimum electric field required for its initiation.
- The development of a probabilistic model reflecting the dispersion parameters that affect the inception of the first corona depending both on the voltage waveform and the HV electrode.
- The study of the spatial and temporal development of the corona discharge, in particular the streamers composing it.

Minimum field required for first corona inception

For a given gap geometry, the initiation of the first corona requires a minimum electric field [5, 6]. This latter is given by the well known *Peek's law* [16] described in section 3.2 of this book.

Dispersion model

Experimental results [6] have shown that corona inception voltages present a dispersion that increases with the rise time of the impulse voltage. To simulate such a statistical distribution, a mathematical model [5, 6] using the concept of 'critical volume' was proposed.

As described previously (see section 3.2), a corona is initiated at a given time when the primary avalanche grows in a region where the electric field is high enough. However, at that time a providential electron must be available. If this electron is close to the HV electrode, the avalanche can reach the electrode before developing to its critical size. If it is too far from the electrode in a region where the geometric electric field is lower than the critical one (given by Peek's law), the ionization coefficient α becomes too weak to allow further development of the avalanche. A volume called the critical volume V_c, bounded by two surfaces is then defined (see section 3.2).

The second step consists in defining a probability function $P_e(t)$ related to the presence of a providential electron within V_c, at a given time t. Between t and $(t + \mathrm{d}t)$, this function can be written as [5, 7]:

$$P_e(t)\mathrm{d}t = \left\{ \int V_c(t) \frac{n_n}{\tau} \mathrm{d}V \right\} \mathrm{d}t, \tag{6.1}$$

where n_n is the density of particles susceptible to producing free electrons as a function of the field E, and τ is their average lifespan.

The probability of first corona inception between t and $(t + \mathrm{d}t)$ can then be deduced:

$$P_i(t)\mathrm{d}t = P_e(t)\mathrm{d}t \exp\left(-\int_0^t P_e(u)\mathrm{d}u \right). \tag{6.2}$$

Consequently, the probability of first corona inception applies as follows:

$$P_c = \int_0^\infty P_i(t)\mathrm{d}t = 1 - \exp\left(-\int_0^t P_e(t)\mathrm{d}t \right). \tag{6.3}$$

However, the experimental results reported in the literature are not enough to determine the rate of production n_n/τ [5, 6].

Corona spherical model (Hepworth, Klewe, Tozer)

The corona spherical model assumes that the growth of a divergent field at the electrode leads to a spherical corona charge centred at the HV electrode, due to the expansion process of the corona with the applied voltage, but is subject to certain limiting conditions due to the properties of the basis of electrical avalanches [17]. The equations used for field calculations are those of two concentric spheres.

The results predicted by the model are quite close to those obtained experimentally, but kept to the minimum of the curve.

6.2.2 Streamer propagation models

Dawson and Winn's model

The propagation of the positive streamer is exclusively related to the activity of its head [18]. The electrons are supposed to be neutralized behind it while the filament connecting the anode to the head of the streamer is not taken into account. These authors assume that a space charge of ions N^+ is located at the head of the streamer in a sphere of radius r_s (figure 6.1).

A providential electron at a distance r_1 from the sphere containing N^+ positive ions, initiates an avalanche which develops under the influence of the electric field due to these ions up to r_2. The applied electric field is neglected. The condition required for the propagation is that the avalanche produces additional N^+ positive ions located in a sphere of radius r_s before reaching the head of the streamer, therefore $r_2 > 2r_s$.

The number of positive ions N^+, created by the avalanche is given by

$$N^+ = \exp\left(\int_{r_2}^{r_1} \alpha(r)\mathrm{d}r\right). \tag{6.4}$$

The radius r_0 of the streamer head is the radius of the avalanche, as follows:

$$r_s^2 = 6\int_{r_2}^{r_1} \frac{D_e}{u_e}\mathrm{d}r \tag{6.5}$$

D_e being the diffusion coefficient and u_e the electron drift velocity.

The starting point of the avalanche r_1 is defined by

$$E(r_1) = E(\alpha = \eta) = \frac{N^+}{4\pi\varepsilon_0 r_1^2} = 22.8\ \mathrm{kV\ cm^{-1}}, \tag{6.6}$$

where η is the attachment coefficient and $E(r_1)$ is the field at which ionization is compensated by the attachment.

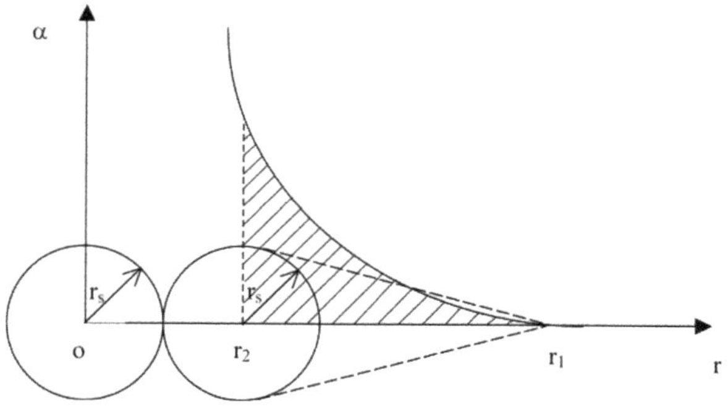

Figure 6.1. Development of an avalanche at a positive streamer head according to the Dawson and Winn model. Reproduced with permission from [5].

From this set of equations, the Dawson and Winn show that stable propagation is possible for $r_s = 2.7 \times 10^{-3}$cm and $N^+ = 10^8$ ions. This value represents the minimum number of particles within the head of the streamer required for auto-propagation without an external applied field.

In the case of a divergent field, these authors determined the maximum elongation of the streamer, considering that the energy loss due to ionization and excitation is compensated during the drift of positive space charge by the input potential energy due to the applied field. The values predicted by this model differ from experimental results. This difference might be due to an improper evaluation of energy losses occurring during the development of an avalanche. An accurate assessment was proposed in a streamer propagation model developed by Gallimberti [6].

Gallimberti model

For the Gallimberti model the same hypothesis is assumed, i.e. a spherical head of the streamer of radius r_s containing N^+ positive charges [6]. In front of the streamer head is a region, referred to as the 'active region' (in contrast to the 'cold area', the streamer channel assumed to be electrically neutral and weakly ionized). Gallimberti [6] assimilates the generated avalanches to a unique equivalent avalanche at P_0 (figure 6.2).

The position of P_0 is derived from the principle of energy conservation and the macroscopic equations of continuity for the electronic density n_e and ionic densities, N^+ and N^-.

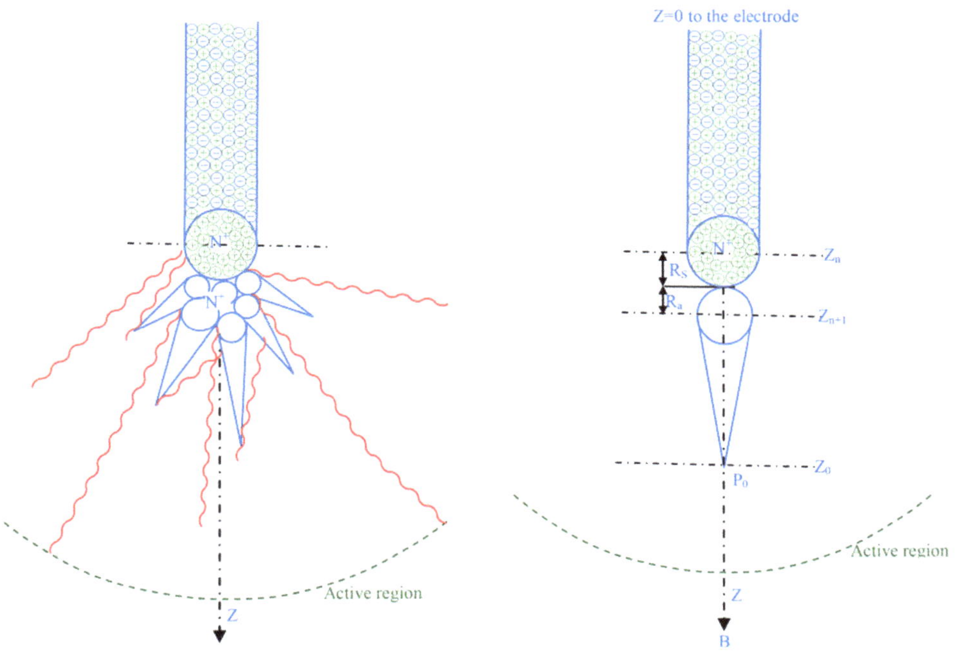

Figure 6.2. Schematic representation of the mechanism of development of a streamer. (A) Diagram showing the trajectories of photons and the series of new avalanches. (B) Simplified diagram of a single avalanche. Reproduced with permission from [5].

The energy balance is written as

$$W_1 = W_g + W_{pot}, \tag{6.7}$$

where W_g represents the energy related to the geometric field E_g

$$W_g = \int_{Z_n}^{Z_{n+1}} \left(eN_eE_g\right)\mathrm{d}z. \tag{6.8}$$

N_e is the linear density along the axis z. W_1 represents the total energy losses including losses due to ionization, attachment, excitation of neutral molecules at different excitation levels, elastic collisions, and the variation of the kinetic energy of the electrons along the course of the avalanche.

The potential energy W_{pot} of the sphere of space charge, taking into account its own energy and the energy of this charge in the field E_g

$$W_{pot} = \frac{0.43(N^+e)^2}{4\pi\varepsilon_0 R_s} + eN^+\int_0^{Z_n} E_g\mathrm{d}z. \tag{6.9}$$

The author showed that

$$W_1 = \beta_1 N_s' E_g, \tag{6.10}$$

with $\beta_l = 0.39 \times 10^{-24}$ cm, N_s' being the charge produced in the avalanche and

$$W_1 = \gamma_1 N_s', \tag{6.11}$$

with $\gamma_l = 49$ eV.

Introducing equation (6.10) into the energy conservation equation, it yields

$$\frac{N_s' - N_s}{N_s} = \frac{\beta_1 E_g - \gamma_1 + \dfrac{0.4N_s e^2}{4\pi\varepsilon_0}\dfrac{(R_s' - R_s)}{R_s^2} + E_g e(R_s + R_s)}{\dfrac{0.4e^2}{4\pi\varepsilon_0}\dfrac{2N_s}{R_s} + eU + \gamma_1 - \beta_1 E_g}. \tag{6.12}$$

Equation (6.12) allows estimation of the charge variation in the streamer head during the propagation. If the initial conditions N_s, R_s, E_g are known as well as the local applied voltage U, the new radius R_s' is calculated as function of N_s and E_g.

The predicted current shape is consistent with measurements. However, the peak values are lower than the experimental ones. This difference is probably due to the large number of streamers and their ramification, a parameter not included in the single avalanche model.

6.2.3 Leader models

Formation and propagation of positive leader

The high conductivity of the leader allows one to distinguish the boundary between its strict channel and the diffusion region (the corona ahead). The evolution of the stem provides the first element of the leader's channel [19]. According to Gallimberti [6], the conductivity of the stem can be expressed, by

$$\sigma = (n_e\mu_e + n_-\mu_- + n_+\mu_+), \tag{6.13}$$

where n_e, n_-, n_+ are, respectively, the densities of electrons, negative ions, and positive ions; μ_e, μ_-, and μ_+ are the corresponding mobilities. Electron density n_e should be much lower than n_+ and n_-; however, considering the high mobility of electrons, their thermal detachment (which occurs at temperatures of about 1500 K) from negative ions and therefore the concomitant increase in their density may lead to an important increase in the stem conductivity; the net positive charge in the stem is therefore rapidly increased. The increase in both stem conductivity and net positive charge density leads to a sharp increase of the electric field around the stem. This may cause a new ionization front, the inception of a second corona, and leader formation [5, 6].

Leader canal models
The channel of an existing section of the leader is assimilated to a cylindrical conductor conveying the current of the leader corona towards the anode. Therefore, the properties of the medium depend on three variables: z, r, and t (figure 6.3).

Several leader models were proposed. Their main difference lies in the existence of LTE [3, 5, 6, 15].

The LTE model
Some simplifying hypotheses are formulated, as follows:

- No radial transport of energy occurs towards the outside of the channel.
- The heated gas mass per unit length of the channel is considered constant during the propagation of the leader.
- The temperature of the channel is uniform along a cross section.
- The gas pressure within the channel is constant.
- The heated gas volume is proportional to the temperature in kelvins.

In the case that ETL applies, the electronic density is calculated from Saha's equation [3, 5, 6],

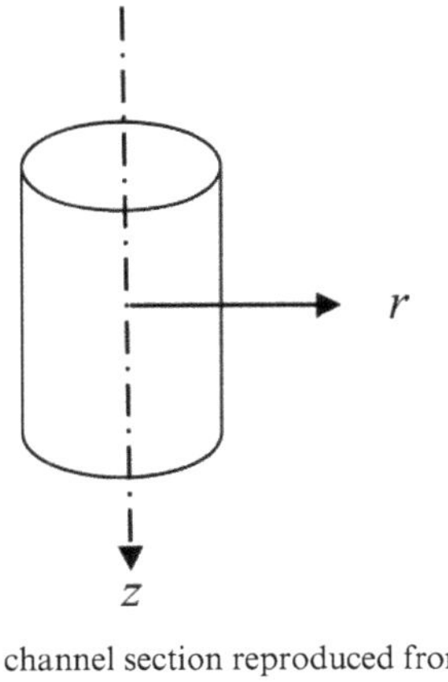

Figure 6.3. A representation of a leader channel section reproduced from [15] with permission of Université de Pau et des Pays de l'Adour, France.

$$n_e = 4.768 \times 10^7 N^{0.5} T^{0.75} \exp(-90420/T)\ (\text{cm}^{-3}), \tag{6.14}$$

where N represents the gas molecular density and T the absolute temperature.

The current flowing in the channel yields

$$I = n_e\, e\, v_e\, \pi a^2 \tag{6.15}$$

where ν_e, electron drift velocity under the leader longitudinal electric field, is given by

$$v_e = 1.9\, E\ T\ (\text{cm}\ s^{-1}), \tag{6.16}$$

and a, the channel radius, increasing linearly from an initial value a_0 such that

$$a = a_0 + r't. \tag{6.17}$$

Numerical applications lead to an electronic density in the range of 10^{12} cm^{-3} and a temperature of 5300 K for given experimental conditions [5]: $I = 0.6$ A, $a = 3$ mm, and $E = 10^3$ V cm^{-1}.

However, the temperature used to determine the electronic density has to be the electronic one, and not that of the gas; the electronic excitement and ionization are probably due to electronic collisions. The electronic temperature should be much higher than 5000 K for the relatively large values of the accepted reduced field E/N values of the leader.

Whirter [5, 20] proposed a criterion for the application of LTE, as follows

$$n_e \geqslant 1.6 \times 10^{12}\ T_e^{1/2}\ X^3\ (\text{cm}^{-3}), \tag{6.18}$$

where X is the largest interval of energy in terms of spectral patterns of present ions. Considering $X = 7$ eV and $T_e = 50000$ K, it yields $n_e = 4 \times 10^{17}$cm^{-3}. Clearly this criterion is not satisfied, and the assumption of LTE should therefore be rejected.

Non-LTE model

When LTE is not attained, the ionization balance can be examined from production rate equations. The channel of the leader is assumed to be a cylindrical filament formed by weakly ionized plasma, where the energy is transferred to the gas by the electric field through collisions between electrons and molecules. The time evolution of the channel's internal conditions (conductivity, temperature, molecular density, gradient of potential, ionization degree, and radius) are established by examining the thermodynamic and hydrodynamic processes resulting from the injected energy.

Since periphery losses are negligible, the leader channel can be considered to expand at a constant mass ($\pi a_0^2 N_0$) (a_0 and N_0 are, respectively, the radius and density of gas molecules at time t_0), the leader portion is formed. The leader current, I, is assumed to be an independent external variable. The electron density is derived from the production rate equations, provided some assumptions are made:

- The channel is a non-attaching medium, i.e. a medium with few negative ions; therefore $n_e = n_+$.
- The detachment rate is offset by the rate of attachment.

It is therefore possible to write for the homogeneous channel:

$$\frac{\mathrm{d}n_e}{\mathrm{d}t} = v_i n_e - a_e n_e n_+, \tag{6.19}$$

where ν_i is the ionization frequency, a_e the recombination coefficient, and n_e the per unit length number of electrons in the channel.

Assuming that the leader channel behaves like a resistance conductor, the potential gradient along the leader is then expressed as a function of the electronic mobility μ_e

$$E = \frac{I}{n_e q \mu_e}. \tag{6.20}$$

Note that, for values of E/N and n_+ attributable to the channel, the time constants of ionization $(1/\nu_i)$ and recombination $(l/a_e n_+)$ are very low in comparison to the duration of leader development. As a result, n_e and n_+ correspond to a balance between the ionization and recombination rates. Therefore

$$n_e = n_+ = \frac{v_i}{a_e}. \tag{6.21}$$

Equation (6.20) may be rewritten as follows:

$$\frac{E}{N} = \frac{1}{N\pi a^2 \frac{1}{a_e}\frac{v_i}{N} N\mu_e q}. \tag{6.22}$$

Since $1/a_e$, ν_i/N, and $N\mu_e$ are only functions of the reduced field E/N while $N\pi a^2$ is constant, E/N appears as an implicit function of I. Therefore the degree of ionization, i, will be

$$i = \frac{n_+}{N} = \frac{1}{a_e}\frac{v_i}{N}. \tag{6.23}$$

Consequently the leader conductivity is

$$\sigma = \frac{1}{a_e}\frac{v_i}{N} N\mu_e q. \tag{6.24}$$

It is solely a function of the leader current.

The variations of E/N, i, σ, and T_e (electron temperature) are shown in figure 6.4 for $a_0 = 1$ mm and $T_0 = 1000$ K. Note in particular that the reduced field E/N is slightly affected by I.

Neglecting radiation and conduction losses, the energy conservation equation can be written as

$$\frac{\mathrm{d}W_t}{\mathrm{d}t} + p\frac{\mathrm{d}(\pi a^2)}{\mathrm{d}t} = EI, \tag{6.25}$$

with $W_t = M\xi$ and $M = mN\pi a^2$, and p is the pressure, ξ the internal energy per unit mass, M the total mass, and W_t, the total energy.

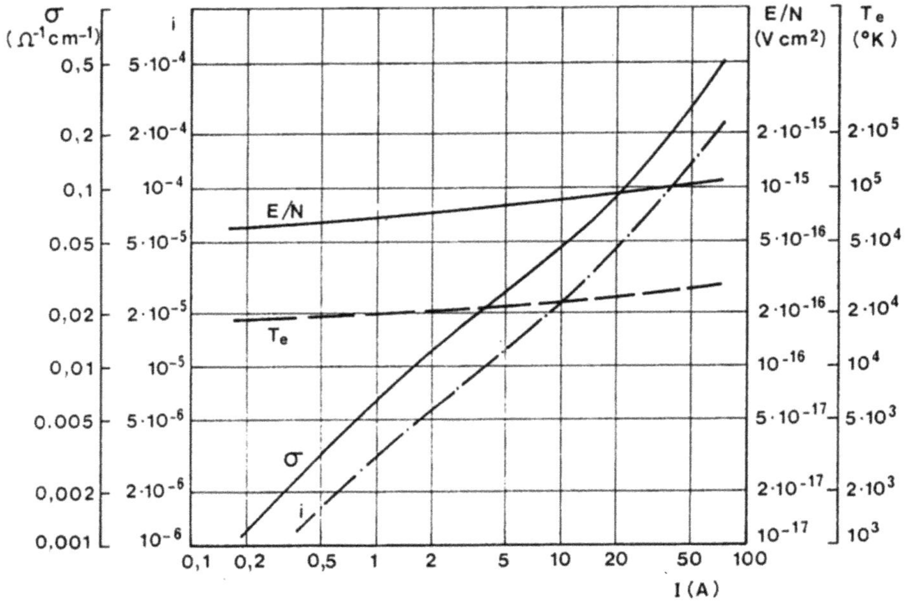

Figure 6.4. Leader channel parameters for $a_0 = 1$ mm and $T_0 = 1000$ K [3]. Figure reprinted with permission from CIGRE, Positive discharges in long air gaps © 1977.

The gas being far from LTE, ξ cannot be treated as a simple state function. This internal energy is calculated from a suitable approximation [3, 5], as follows:

$$\xi = \frac{1}{m}\frac{kT}{\gamma - 1} = \frac{1}{\gamma - 1}\frac{p}{mn}, \tag{6.26}$$

where γ is the ratio of specific heats at constant volume and constant pressure. Experimental results and theoretical simulations [6, 21] show that the channel expands at constant pressure with a subsonic velocity, the pressure p inside the channel can be assumed to be equal to atmospheric pressure p_0. Assuming that the gas in the channel behaves like an ideal gas and the degree of ionization is much lower than 1, the state equation is written as

$$p = NkT, \tag{6.27}$$

where k is Boltzman's constant. Assuming that γ is constant, equation (6.25) can be rewritten, as follows, taking into account the mass conservation

$$\frac{p_0}{\gamma - 1}\frac{\mathrm{d}(\pi a^2)}{\mathrm{d}t} + p_0\frac{\mathrm{d}(\pi a^2)}{\mathrm{d}t}. \tag{6.28}$$

The power injected reads

$$EI = \frac{E}{N}I\frac{N_0\pi a_0^2}{\pi a^2}. \tag{6.29}$$

Given that E/N is almost constant [3, 21], equation (6.20) is combined with equation (6.29) to provide the channel cross section at time t and position x

$$(\pi r^2)_{x,t} = \sqrt{(\pi r_0^2)^2 + 2\frac{\gamma - 1}{\gamma}\frac{E_L}{N}\frac{\pi r_0^2}{k\tau_0}Q(t)}, \tag{6.30}$$

t_0 is the moment when the head of the leader passes position x and where the portion of the channel is formed.

6.2.4 Self-consistent models

The models presented in the previous sections describe a specific physical interpretation of the phenomena observed during the development of an air gap discharge and allow us to improve our knowledge of its internal parameters, which are inaccessible to measurement. However, each model describes only part of the phenomenon, and individually, they may not report the complete discharge phenomena resulting from the inter-correlated phases. In this section, some of the available self-consistent models, which take into account the main phases of the discharge, are presented.

A self-consistent model may be considered as satisfactory if the following conditions are satisfied:

- It should be based on a realistic representation of the discharge.
- It should allow for a prediction the main parameters of the discharge and of their dynamics according to the applied voltage characteristics, the gap geometry, and some atmospheric parameters.
- It should consider the random aspect of a discharge and allow for prediction of the distribution of the main statistical parameters.

Hutzler and Hutzler's model

The analytical model developed by Hutzler and Hutzler [5, 11] consists in assuming the tip of the leader to be an equivalent electrode of radius R moving in the gap.

If the electric field E_p at this electrode surface is sufficient (higher than a critical field E_c) to keep the corona effect alive, the leader channel can propagate through the gap. Otherwise, the leader channel progressively decays, unless the applied voltage characteristics are able to re-activate it. The voltage applied to the uncrossed gap is the potential U_j at the tip of the leader of axial length L_z. The field E_p is estimated using a hyperboloid approximation [11],

$$E_p = \frac{2U_j}{R_e \ln 4\left(\dfrac{D - L_z}{R_e}\right)}, \tag{6.31}$$

where D is the gap length and R_e is the radius of the equivalent hyperboloid electrode.

The voltage at the tip of the leader can be determined from the applied voltage $U_c(t)$,

$$U_j = U_c(t) - \Delta U, \tag{6.32}$$

where the voltage drop, ΔU, along the leader channel is

$$\Delta U = \int_0^L g(M,\ t)\mathrm{d}t, \tag{6.33}$$

where L is the length of the leader at t, and $g(M, t)$ the electric field along the leader channel of length L; M being the position of the leader tip in the gap.

The authors assume the leader to be in an ionized channel where the energy imbalance between different types of particles slowly decreases. The leader channel, therefore, may be considered as arc plasma. The gradient g is calculated by

$$\frac{\mathrm{d}g}{g_\infty(I) - g} = \frac{\mathrm{d}\tau}{\varpi}. \tag{6.34}$$

τ is the age of the considered leader portion, ϖ is the leader aging time constant, i.e. velocity at which g tends towards the asymptotic value $g_\infty(I) = AI^{-\alpha}$ with $A = 0.25 \times 10^5$ and $\alpha = 0.9$.

An empirical expression for R_e was proposed in the following equation:

$$R_e = \frac{U_j}{U_{j0}} \frac{0.25}{1 + \dfrac{3.5}{L}} + R_i \exp\left(\frac{L_z}{-13.6R_i}\right), \tag{6.35}$$

where R_i, L, and L_z are, respectively, the initial equivalent radius after the development of the first corona, and the leader's effective and axial lengths, while U_{j0} is the voltage to be applied to the uncrossed gap to render the field on the equivalent electrode surface equal to the critical field E_c, given by [5, 11]

$$U_{j0} = \frac{Ec_0 R_x}{2} \ln 4\left(\frac{D - L_z}{R_x}\right), \tag{6.36}$$

where

$$Ec_0 = E_0\left(1 + \frac{0.0436}{\sqrt{\delta R_x}}\right), \tag{6.37}$$

$$R_x = \frac{0.25}{1 + \dfrac{3.5}{L}} + R_i \exp\left(\frac{L_z}{-13.6R_i}\right). \tag{6.38}$$

Regardless of the computation method, it is acknowledged that the leader propagates when the electric field at its tip E_p is higher than a critical field E_c deduced from Peek's law.

The field depends on the electrode shape and varies in the range 30–40 kV cm^{-1} for typical electrodes. For a long time, Peek's law (described in chapter 3), has been used in overhead network problems. A generalization of this law [5] has been proposed for the case of electrodes with spherical symmetry. The field E_c (kV cm^{-1}) is expressed as function of R_e (cm)

$$E_c = 31\left(1 + \frac{0.436}{\sqrt{R_e}}\right). \tag{6.39}$$

Intuitively, the higher the electric field, the higher corona current is. Since the leader channel is fed by the corona current, the authors modified a relationship proposed by Townsend to correlate the leader current to the electric field

$$I = hE_p(E_p - E_c). \tag{6.40}$$

h is a constant taken at 0.3×10^{-11} MKSA. This value provides values close to experimental ones.

The leader propagation velocity is strictly related to the corona current [5, 11]

$$v = cI^{0.3}. \tag{6.41}$$

With $c = 1.75 \times 10^4$ MKSA, experimental results were best fitted.

It is known that the hazardous propagation of the leader and dispersion of the inception voltage are responsible for dispersion of voltages and, at times, to flashover. Hutzler and Hutzler [11] have established a probabilistic law which seems to be physically plausible. The erratic propagation is taken into account with a law of probability characterized by a dispersion parameter a. The probabilistic density $P(\varphi)$ of angle φ is

$$P(\varphi) = \frac{\sin\varphi}{a}\exp\left(-\frac{1-\cos\varphi}{a}\right), \tag{6.42}$$

where a is a constant related to the variation in space and time of the field at the end of the leader.

The final jump is assumed to start when the average electric field between the leader equivalent electrode and the cathode reached the value g_s ($g_s = 4$ kV cm^{-1}) of the electric field along the streamers, i.e.

$$\frac{U_j}{D - L_z} \geqslant g_s \tag{6.43}$$

The set of relationships governing the model is solved provided the geometric characteristics of the gap and the voltage waveform are known. The resolution is then determined according to the chart in figure 6.5 [5, 11].

The main criticisms of this model can be summarized in four points:

- Most of the relationships used are based on empiricism or barely acceptable assumptions.
- A relationship based on arc theory is used to account for the leader gradient.
- While the amplitude of the predicted leader current agrees quite well with experimental data, their waveforms are very different.
- The proposed model does not simulate the return stroke phase.

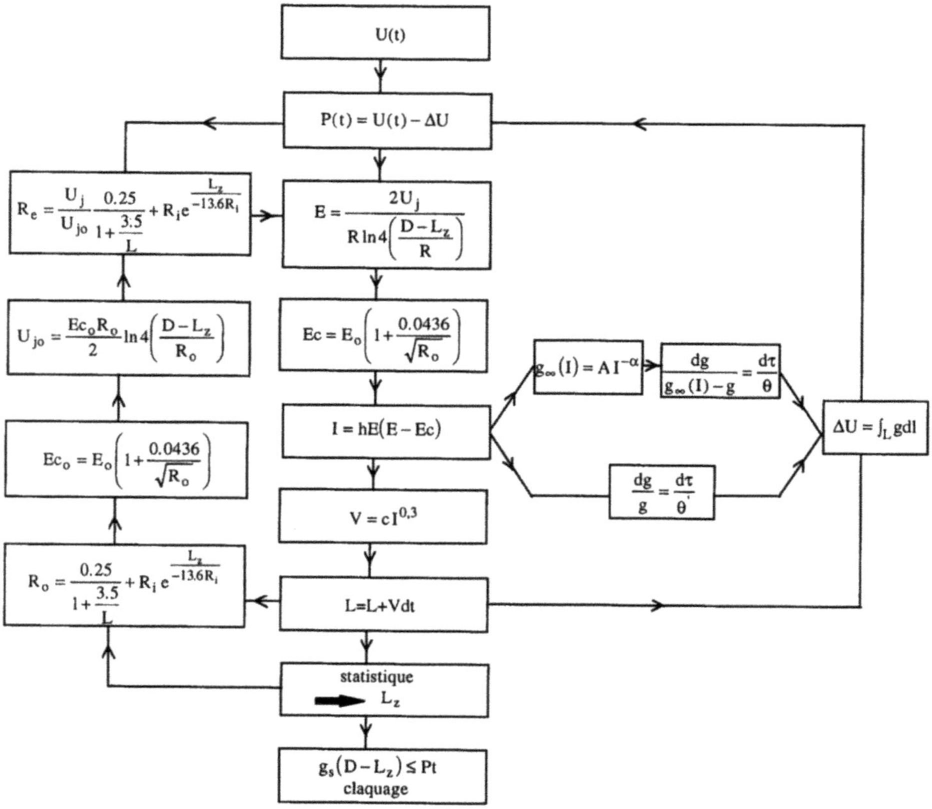

Figure 6.5. General flowchart of Hutzler and Hutzler's model reproduced with permission from [11].

Ortega's model

This model was established for large air gaps subjected to voltage waves of pure or oscillating bi-exponential shapes [7]. The general idea consists in modelling the discharge propagation by a discreet injection of charge in the gap.

From experimental measurements, the author established a relation between the apparent charge and the voltage level at the three different phases of the discharge development (first corona, first oscillation, final jump):

- first corona : $t = t_i$ $\quad Q_i = k_i U(t_i)(U(t_i) - U_0)$
- first oscillation: $t = t_p$ $\quad Q_p = k_p U(t_p)(U(t_p) - U_0)$
- final jump: $t = t_{\mathrm{sf}}$ $\quad Q_{sf} = k_{sf} U(t_{sf})(U(t_{sf}) - U_{sf})$

with

$$U_0 = \frac{1}{2} R_{\mathrm{p}} E_0 \log\left(\frac{4D}{R_{\mathrm{p}}}\right)\left(1 + \frac{0.0436}{\sqrt{R_{\mathrm{p}}}}\right). \tag{6.44}$$

U_0 is the theoretical corona inception voltage, deduced from Peek's law for the point–plane geometry [7]; k_{sf}, k_{p}, and k_{i} are parameters that depend on experimental conditions.

R_p is the HV electrode's radius; D is the gap length, and E_0 is the critical field at which the ionization phenomena become significant.

Two relationships that allow for the description of the injected charge versus time were proposed:

- The first one $Q_1(t) = k_1(t)U(t)(U(t)-U_0)$ is related to discharge propagation under optimal conditions (ideal propagation).
- The second $Q_2(t) = k_2(t)U(t)(U(t)-U_i)$ is related to the true conditions of development of the discharge (true propagation). This second relationship takes into account the statistical time lag and the discontinuity of the charge injection into the gap with time.

Based on experimental results, $k_1(t)$ and $k_2(t)$ increase with time. Ortega [7] then proposed both

$$k_1(t) = k_i + k_c L_c(t) \tag{6.45}$$

$$k_2(t) = k_i + k L_c(t). \tag{6.46}$$

$a_1 = k/k_c$ is defined as a propagation quality coefficient. It reflects the excess of injected charge consumption per unit axial length relative to the true propagation case. The following coefficients were determined experimentally:

- $k_i = 40\ \mu\text{C MV}^{-2}$, which characterizes the first corona.
- $k_c = 12\mu\text{C MV}^{-2}$, which characterizes the channel elongation in the case of the ideal propagation.

The basic principles of the charge injection system modelling are shown schematically in figure 6.6.

At time t_i, when the first corona is initiated, the injected charge ΔQ_i is given according to $Q_1(t_i)$ and no additional charge appears in the gap.

At time t_1, we have $Q_2(t) = Q_1(t_i)$, where the additional charge injected ΔQ_1 is given by the equation $L_z(t) = Q_1(t)/\lambda$, where λ is the average charge per unit length deduced experimentally. As long as $Q_1(t) > Q_2(t)$, the above-mentioned operation is repeated at time intervals dt until the final jump phase.

The development of the discharge is therefore characterized by the following equation derived from the model of Baldo and Pesavento [7],

$$\mathrm{d}L_z(t) = \left[k_1(E_s - E_0) + k_2\frac{\mathrm{d}U}{\mathrm{d}t}\right]\mathrm{d}t. \tag{6.47}$$

k_1 and k_2 are determined from experiments. The general flowchart of the modelling is presented in figure 6.7.

The final jump starts as soon as the axial length of the leader L_z is higher or equal to the so-called 'final jump' L_{fj} deducted from the Garcia and Hutzler law [22], as below.

$$L_{fj} = D - \frac{U_c(t) - 0.15}{0.44}\ (\text{MV m}) \tag{6.48}$$

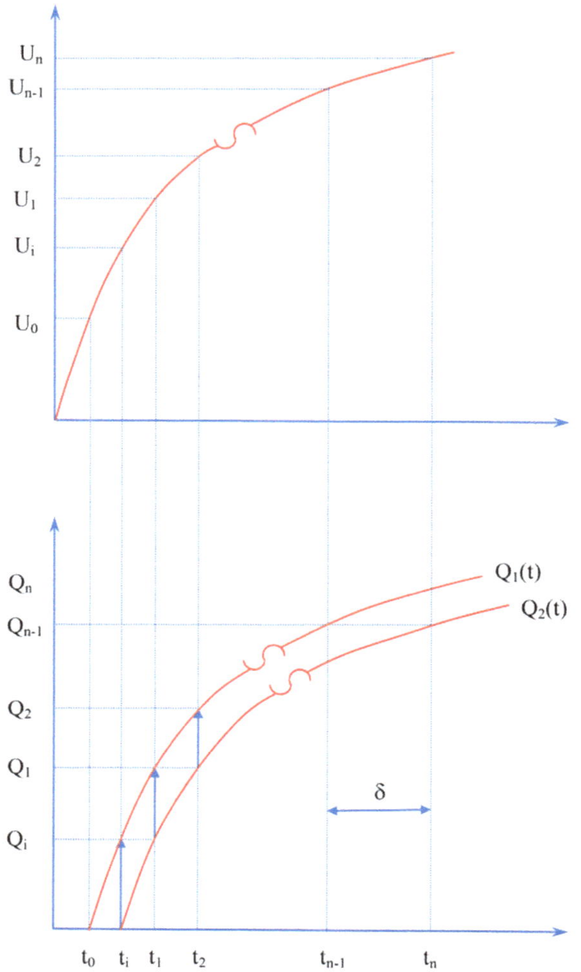

Figure 6.6. Basic principles of charge injection system modelling reproduced with permission from [7].

The main criticisms of this model can be summarized in the following five points:

- Most of the relationships used are based on empiricism.
- The model requires experimental data relative to a given electrode configuration for the complete simulation.
- The physical properties of the discharge, such as the conductivity of the leader channel or the evolution of its radius, are not at all taken into account in the calculations.
- Parameters relevant to atmospheric conditions are not included.
- The proposed model does not simulate the return stroke phase.

Lemke's model

This model is based on the observations of the spatial structure of the discharge [23]; it enables one to predict the breakdown voltage in a divergent field for gaps lower

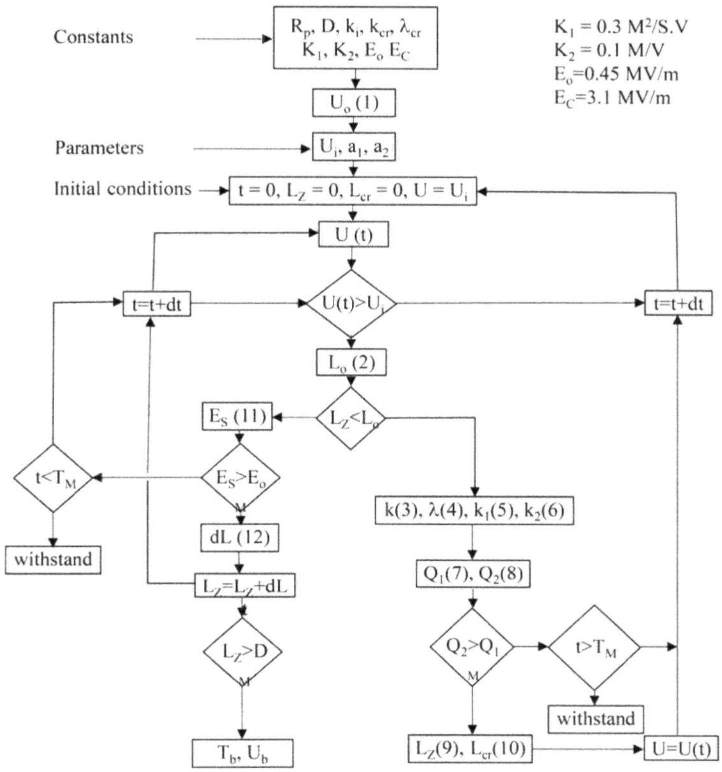

(1) $U_0 = \frac{1}{2} R_p E_C \log\left(\frac{4D}{R_p}\right)\left(1 + \frac{0.0436}{\sqrt{R_p}}\right)$	(7) $Q_1(t) = k_1(t) U(t)(U(t) - U_0)$
(2) $L_0 = D - \left(\frac{U(t) - 015}{0.44}\right)$	(8) $Q_2(t) = k_2(t) U(t)(U(t) - U_i)$
(3) $k = a_1 k_{cr}$	(9) $L_z = Q_1(t) / \lambda$
(4) $\lambda = a_2 \lambda_{cr}$	(10) $L_{cr} = Q_1(t) / \lambda_{cr}$
(5) $k_1(t) = k_i + k_{cr} L_{cr}(t)$	(11) $E_S = (U(t) - 0.15) / (D - L_z)$
(6) $k_2(t) = k_i + k L_{cr}(t)$	(12) $dL = K_1(E_S - E_0)dt + K_2 dU / dt$
	$T_M = 200 \mu s$ or $T_M = 400 \mu s$
	$dt = 0.2 \mu s$ or $dt = 0.4 \mu s$

Figure 6.7. General flowchart of Ortega's model reproduced with permission from [7].

than 10 m. The interesting aspect of this model is that it takes into account, to a large extent, the physical phenomena of the discharge and that it better describes the state of the discharge from the moment when the final jump starts.

However this model presents some weaknesses as it is incapable of predicting the behaviour of curves U and the expression of the gradient in the proposed leader leads to very low (weak) values. Also during its development, the leader is characterized by a velocity and a constant current, suggesting that the voltage at the head of the leader is practically constant. Thus it is difficult to reconcile this hypothesis with the dependency of the streamer length to that of the leader.

Carrare and Thione's model

This model is based on miscellaneous observations and analyses of the discharge structure and the breakdown voltage of rod–plane, sphere–plane, torus–plane, and wire–plane spark gaps [24]. It considers that the appearance of the leader channel is conditioned by the creation of a critical space charge and that the value of the voltage at the fore of the leader is constant.

The proposed model enables the determination of the breakdown voltage of various insulating systems with sufficient precision for practical objectives, where the electrodes have spherical or cylindrical forms.

Bondiou and Gallimberti's model

This is a physical model that takes into account the electrostatic, hydrodynamic, and thermodynamic processes occurring continually in the development of the discharge [25]. These considerations enable it to simulate the initiation and development of the discharge up to the final jump. The simulation of the corona and the phases of the leader is coupled with analytical calculations of the electric field at the electrodes, in the leader channel, and the space charge injected into the gap. The only data of the model are the electrodes geometry and the applied voltage. This model also enables us to simulate the behaviour of the discharge with or without over-voltages, perform the analysis of the conditions of stable propagation of the leader, as well as determine very important parameters such as the voltage U_{50}. Unfortunately, the equations that we have to handle in this model are quite challenging and the shapes of the currents of the leader differ notably from those reported in the literature. Note also that this model does not include the return stroke phase.

Aleksandrov's model

Aleksandrov proposed a model enabling one to compute the breakdown voltage of very long air gaps (some km) in a wire–plane system [19]. The obtained values are higher than those obtained experimentally and do not present a decrease of the dielectric strength with the increase of the electrodes gaps, even with significantly large values.

Jones' model

This model, also called the 'column model', assumes the discharge to the growth of a conductive column of constant radius into the gap [9]. The model supposes a cylindrical distribution of charge. The progress of the leader is connected to a field criterion. The conductive part of the column is assumed to be the arc. This model enables us to obtain the U curve. The obtained results are rather close to those obtained experimentally. Among the criticisms of this model are: it assumes the radius of the leader channel should be a constant; the critical field is not calculable but approximately obtained; the conductive phase of the leader is assimilated to the arc; and the current of the discharge is considered as a constant.

Corona spherical model (Hepworth, Klewe, Tozer)

This model assumes that the growth of a divergent field at the electrode leads to spherical corona charges centred at the HV electrode, due to the expansion process of the corona with the applied voltage. However, it is also subject to certain limitative conditions due to the properties of the basis of electrical avalanches. The equations used for field calculations are those of two concentric spheres. The results predicted by the model are quite close to those obtained experimentally, but they are kept to the minimum of the curve.

6.3 Negative discharge models

6.3.1 Transition corona-streamer–leader

At the base of the corona, the electrons only provide kinetic energy to gas molecules. It was shown that in this case, the increase in the gas temperature can be more than 5000 K [19]. The increase in the conductivity of the gas is then explained by the creation of charges by thermal ionization.

During the development of corona, a current is injected at the root of the streamers. The corresponding electrical energy is transferred mainly in the vibrational form through heavier molecules by electron–neutral gas collisions.

These molecules, by de-energizing, gradually restore this energy to the medium as heat. The relaxation time of the vibrational energy levels thus determines the temporal evolution of the temperature. Beyond 1500 K, the negative ions lose their excess electrons. We then witness a sharp increase in conductivity at the root of the streamers, also called the 'stem'. This stem, similar to a conductive projection (protuberance) on the surface of the electrode, allows us to achieve the electrical conditions for the formation of the second corona.

After the development of one or more coronas, there is the presence in the gap of one or more bright nuclei whose activity is more or less periodic, called the 'space stem'. The latter stands for the initiation point of a system consisting of negative and positive streamers propagating in opposite directions. The system formed by the space stem and streamers constitutes the 'pilot' [26].

The created pilot propagates into the gap towards the anode as a bidirectional corona on a fictitious electrode. It emits negative streamers that propagate in a virgin environment. Two models have been proposed to explain the formation of this pilot.

6.3.2 The spatial and negative electrode leaders

Experimental observation showed that a leader has a filamentary aspect, the trajectory of which is irregular; it often propagates in a direction making a considerable angle with respect to the direction of the applied field. Its propagation velocity depends on the geometry of the space and the applied voltage [27]. The brightness of the leader is continuous, provided that the rate of increase of the applied voltage is not too low or under certain conditions where elongation and sudden brightness of the channel occur. Such phenomena are known as re-illuminations (restrikes). These are high speed discrete increases of the length of the leader accompanied by a large pulse of the corona and a brief increase in the brightness of the leader and the discharge current.

The corona of the leader is a large (voluminous) discharge showing a filamentary structure which ionizes the air at the front of the leader channel. This is the mechanism by which the current flowing in the discharge which develops is maintained and changes the electric field distribution in space [6].

The filaments of corona have a common origin at the electrode from which every streamer issues. This meeting point of streamers is called the 'cathode stem'. Its evolution in a negative leader is explained in the same way as for the positive leader in the case of a point of negative polarity. Schlieren records [28] did not reveal any difference between these two types of leader. That is why the models of leaders established in the case of a positive point are applied to negative polarity.

During the final phase of the jump that leads to the arc, the channel leader comes in three forms (figure 4.2(a)) according to the observations of Gorin and Shkilev [26]:

- A channel, called a 'negative leader' that grows from the cathode.
- A channel that develops from a point in the gap towards the two electrodes that may cause the lengthening of a negative leader with a re-illumination, called a 'spatial leader'.
- A channel that develops during the final jump, from a point on the plane toward the cathode, called a 'positive leader'.

The phase that includes the negative leader and one or more spatial leaders is called the 'phase leader' and is analogous to positive polarity.

6.3.3 Models of leaders

The main models of leaders, established in positive polarity and transferable under certain conditions to negative polarity are listed below.

Hutzler and Kleimaier's model

A mechanism that could explain the propagation and emission of positive and negative streamers from a space stem was proposed by Hutzler and Kleimaier [29], assuming that there are two kinds of negative streamers. The former are supplied with electrons from the cathode, which would fully neutralize the positive space charge deposited by successive avalanches. The latter are isolated from the cathode so that they could not provide them with electrons that can neutralize the accumulated positive space charge. The space stem is plasma resulting from a negative streamer of the second type, i.e. a streamer with the positive space charge that cannot be completely neutralized by the available electrons. As a result, the space stem is a dipole with an excess of negative charges. The separation of these charges is provided by the fields E_0 wherein the dipole is confined. The spatial stem is replaced by two spheres of radius r_p and r_n containing charges uniformly distributed Q_p and Q_n (figure 6.8).

After calculating the electric fields E_A an E_B at the positive and negative ends of the dipole, and considering how to simplify the calculations, with the radii of spheres containing Q_p and Q_n identical ($r_p = r_n = r$), we obtain the following expressions:

$$4\pi\varepsilon_0 E_0 rd = Q_p + Q_n \tag{6.49}$$

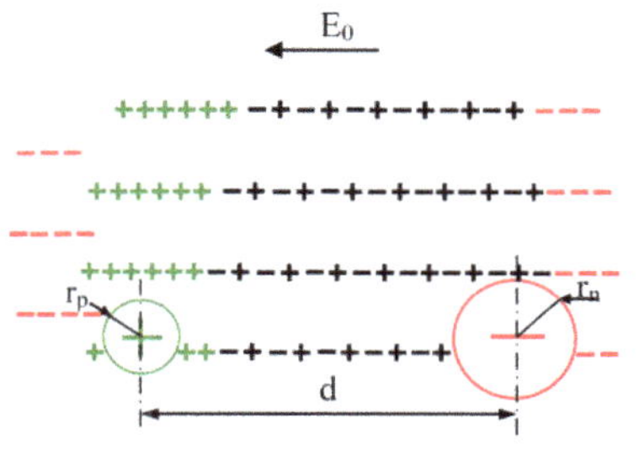

Figure 6.8. Schematic representation of the space stem.

$$E_A = E_0 + \frac{Q_n}{4\pi\varepsilon_0 r^2} \tag{6.50}$$

$$E_B = E_0 + \frac{Q_p}{4\pi\varepsilon_0 r^2}. \tag{6.51}$$

If the charges Q_p and Q_n are such that the E_A field exceeds the critical field E_c, the trigger conditions of a negative streamer are met. Since the space stem is able to provide the electrons needed to neutralize the positive space charge deposited by the negative streamer, this is the first streamer type. A negative charge $-Q_n$ is transferred from the stem to a more advanced position in the gap. If the total negative charge Q_n decreases in the stem, this translates into an increase of the E_B field. As a result, this field may exceed the value in turn the critical field E_c. Therefore a positive streamer may be triggered.

The created streamer is equivalent to an injection of negative charge $-q_p$ in the stem and the apparent drift of a positive charge Q_p towards the anode. Equilibrium is re-established so that the increase of Q_n can give a new emission sequence. On recordings from an image converter, it has been observed [27] that a negative streamer is followed by about three positive streamers. Hutzler acknowledged that the charge, Q_n, associated with negative streamers is about three times the charge q_p associated with positive streamers. Consequently, two main problems arise with the Hutzler model:

- Increasing the E_A field is insufficient to attain the critical value E_c.
- Reduction of the E_B field is insufficient to prevent the formation of new positive streamers.

Gallimberti's model

As in the Hutzler and Kleimaier model [29], the goal is to find mechanisms that can lead to the establishment of a dipolar structure [30]. During the development of the first corona, a front of net charge forms before the head of the streamer filament. The increase in net charge at the end of the corona, due to the drift of ions towards the cathode, reduces the value of the field behind the front of the discharge, and increases it between the front and the anode. When the electric field is sufficiently increased, this area may allow the emergence of a new negative corona. The current injected by the new corona very quickly lowers the voltage at the end of the streamer. Thus, the new voltage distribution gives rise to a positive field sufficient to allow the development of a positive corona; the latter propagates towards the cathode [30]. A bipolar system is so created.

6.4 Fractal models of long discharges

Many theories and models such as field mapping, and fractal and percolation models [31–36] have been proposed to simulate the chaotic nature displayed by the discharge channel [37]. These models [31–36] are based on the fractal approach proposed by Niemeyer *et al* [31] for the dielectric breakdown and the use of the stochastic model proposed by Sawada *et al* [38] to take into account the tendency of discharges to branch into random patterns.

The fractal approach was first used for the modelling of lightning discharges to make predictions of the probability of lightning strikes to practical structures (isolated objects, earthed objects, etc) [39–42]. It has been used in particular to describe the effects of tortuosity and branching of the lightning channel by considering that the growth of the discharge is stochastic in time, and the probability of streamer formation is proportional to a certain function of the local electric field [33]. Some of specific estimates include the probability of strikes as function of interception angle, predictions of the strike points on structures, and the probability of 'side strikes' to tall structures. The model has also been used to investigate lightning discharges between the tops of thunderclouds and the ionosphere [43].

The main parameters describing the fractal model are: the critical electric field intensity for propagation of a leader, the electric field intensity in the leader channel, the initial potential of the lightning leader tip (or thundercloud potential), and the distance between the leader tip and the earthed structure [32, 40, 41, 44]. The electric field intensity is determined from the solution of the Poisson equation, i.e. taking space charge into account. The development of the upward discharge from the structure is determined by a condition of leader propagation [45]. Unlike methods which assume the leader propagates in the direction of maximum electric field intensity, the model allows the leader to propagate in a direction which may deviate from the electric field lines. The probabilistic nature of the leader orientation is applied up until the final jump phase on its propagation. At every step of the development of the discharge, the potentials at specified points are determined via the use of the iteration method for the solution of the Poisson equation. The procedure is repeated until the probabilities in all directions become 0.

Note also that some authors have developed numerical models using finite element methods to simulate discharges in air. These models are based on the hydrodynamic drift–diffusion approximation. A set of continuity equations accounting for the movement, generation, and loss of charge carriers (electrons, positive and negative ions) is coupled with Poisson's equation to take into account the effect of space on the electric field [46, 47].

References

[1] Les Renardières Group 1972 *Electra* **23** 53
[2] Les Renardières Group 1974 *Electra* **35** 49
[3] Les Renardières Group 1977 *Electra* **53** 31
[4] Meek J M and Craggs J D 1978 *Electrical Breakdown of Gases* (New York: Wiley)
[5] Le Roy G, Gary C, Hutzler B, Lalot J and Dubanton C 1984 *Les Propriétés diélectriques de l'Air et les Trés Hautes Tensions* (*Collection de la Direction des Études et Recherches d'Électricité de France* vol 51) (Paris: Eyrolles) p 478
[6] Gallimberti I 1979 *J. Physique Coll.* **40** 193
[7] Ortéga P 1992 Comportement diélectrique des grands intervalles d'air soumis à des ondes de tension de polarité positive ou négative *PhD Thesis* Université de Pau et des Pays de l'Adour, France
[8] Rühling F, Heilbronner F, Ortega P, Domens P, Dupuy J and Gilbert A 1991 Long air gap discharges under non-standard impulse voltages. Part 1: Comportement diélectrique: experimental set-up and effect of superimposed oscillations *7th Int. Symp. High Voltage Engineering* (*Dresden, Germany, 26–30 August*) p 101
[9] Jones B 1973 *Phil. Trans. R. Soc. Lond. A* **275** 165
[10] Klewe R C, Waters R T and Jones B 1974 Breakdown phenomena in long air gaps under switching impulse conditions *IEEE Summer Meeting* paper 74-CH 0910 PWR, chapter 4
[11] Hutzler B and Hutzler D 1982 *Bull. Dir. Etud. Rech.* **4** 11
[12] Dellera L and Garbagnati E 1990 *IEEE Trans. Power Deliv.* **5** 2009
[13] Goelian N, Lalande P, Bondiou-Clergerie A, Bacchiega G L, Gazzani A and Gallimberti I 1997 *J. Phys. D: Appl. Phys.* **30** 2441
[14] Reather H 1964 *Electron Avalanches and Breakdown in Gases* (London: Butterworth) p 390
[15] Domens P 1987 Contribution à l'étude des décharges électriques dans les grands intervalles d'air *PhD Thesis* Université de Pau et des Pays de l'Adour, France
[16] Peek F W 1929 *Dielectric Phenomena in High Voltage Engineering* (New York: McGraw-Hill)
[17] Herpworth J K, Klewe R C and Tozer B A 1972 *J. Phys. D: Appl. Phys.* **5** 730
[18] Dawson G A and Winn W P 1965 Z *Phys.* **183** 159
[19] Aleksandrov G N 1966 *Sov. Phys.—Tech. Phys.* **10** 948
[20] Berger K 1967 *J. Franklin Inst.* **283** 478
[21] Les Renardières Group 1986 *Proc. IEEE A* **133** 395
[22] Garcia H and Hutzler B 1974 Electrical breakdown in long air gaps. The final jump *IEEE Third Conf. on Gas Discharges* (*London, UK, 9–12 September*) p 206
[23] Lemke E 1967 Durchschlagmechanisms und Schlagweite-Durchschlagspannungs-Kennlinien von inhomogenen Luftfunkenstrecken bei Schaltspannungen *PhD Thesis* Technical University of Dresden
[24] Carrara G and Thione L 1979 *IEEE Trans. Power Appar. Syst.* **95** 512
[25] Bondiou A and Gallimberti I 1994 *J. Phys. D: Appl. Phys.* **27** 1252
[26] Gorin B N and Shkilev A V 1976 *Elektrichestvo* **6** 31
[27] Les Renardières Group 1981 *Electra* **74** 67
[28] Gibert A, Domens P, Hutzler B, Riu J P and Edlinger F 1981 *RGE Rev. Gen. Electr.* **5** 399
[29] Hutzler B and Kleimaier M 1981 *Electra* **74** 117
[30] Gallimberti I 1972 Mathematical model for streamer model in non-uniform gaps *Padova University Report* Upee 72/03
[31] Niemeyer L, Pietronero L and Wiesmann H J 1984 *Phys. Rev. Lett.* **52** 1033

[32] Petrov N I and Petrova G N 1995 *Tech. Phys.* **40** 427

[33] Petrov N I, Petrova G N and D'Alessandro F 2003 *IEEE Trans. Dielectr. Electr. Insul.* **10** 641

[34] Femia N, Niemeyer L and Tucci V 1993 *J. Phys. D: Appl. Phys.* **26** 619

[35] Pasko V P, Inan U S and Bell T F 2000 *Geophys. Res. Lett.* **27** 497

[36] Mansell E R, MacGorman D R, Ziegler C L and Straka J M 2002 *J. Geophys. Res.* **107** 4075

[37] Petrov N I and Petrova G N 1993 *Tech. Phys.* **38** 287

[38] Sawada Y, Ohta S, Yamazaki M and Honjo H 1982 *Phys. Rev. A* **26** 3557

[39] Kawasaki Z I, Nakano M, Takeuti T and Hasegawa T 1988 Numerical simulations of Lightning by means of the leader propagation model *Proc. 8th Int. Conf. on Atmospheric Electricity* (*Uppsala, Sweden 13–16 June*) p 483

[40] Petrov N and Perova G N 1992 Modeling of lightning channel branching and bending *Proc. 9th Int. Conf. on Atmospheric Electricity* vol 3 p 675

[41] Petrov N and Perova G N 1993 Modeling of the trajectory of leader discharge development *Proc. 8th Int. Symp. High Voltage Engineering* (*Yokohama, Japan, 23–27 August*) vol 4 p 101

[42] Takeuti T, Hashimoto T and Takagi N 1993 *J. Atmos. Electr.* **13** 9

[43] Petrov N and Petrova G N 1999 *Zh. Tech. Phys.* **69** 134

[44] Petrova G N 1998 Lightning stroke simulation by means of fractal approach: attractive and protective structures *Proc. 24th Int. Conf. on Lightning Protection* (*Birmingham, UK, 14–18 September*) p 478

[45] Petrov N and Waters R T 1995 *Proc. R. Soc. Lond. A* **450** 589

[46] Tran T N, Golosnoy I O, Lewin P L and Georghiou G E 2011 *J. Phys. D: Appl. Phys.* **44** 15203

[47] Bedja A 2006 Modélisation de l'attachement de la décharge de foudre sur un paratonnerre, prenant en compte les charges d'espace *PhD Thesis* Ecole Centrale de Lyon, University of Lyon

Chapter 7

Modelling the positive discharge in long air gaps

7.1 Introduction

Discharge phenomena may be deemed a blend of a variety of aspects, including chemical, mechanical, thermal, and electrical properties, as described in the previous chapters. The complexity of these phenomena and especially the gas discharge physics is such that it is difficult to establish mathematical models that take into account all the processes implicated in the discharge. However, the use of some simplifying assumptions renders it possible to model some phases of the discharge. Thus, several models have been presented over the years to explain the mechanism of a propagating discharge in air [1–12], although not all the pertinent aspects of this very dynamic phenomenon have yet been discovered. Most of these models are each limited to a single phase and thus cannot be combined to form a complete model able to predict the entire temporal evolution of the discharge for a given gap and voltage.

The development of a realistic discharge model requires the adoption of a dynamic approach based on the physical processes that constitute the phenomena. Furthermore, the mathematical equations used in such a model must describe these processes so that the model performance is likely to be accurate over a wide range of experimental conditions.

In this chapter, a dynamic model is presented that has been developed over the years by the authors. It simulates electric discharge inception and propagation in long air gaps under a positive impulse voltage, using an electrical circuit concept. The associated physical phenomena are translated into mathematical equations, to be used in the electrical circuit. This model takes into account important parameters, such as electrode geometry, impulse voltage shape, and some atmospheric conditions. The dynamic procedure includes major parameters that influence the phenomenon. These parameters include a field-based leader propagation criterion and leader velocity together with its erratic propagation.

Despite the importance of this model, which may be considered the 'first complete model' developed for predicting the breakdown process of long air gaps, improvements

doi:10.1088/978-0-7503-1236-3ch7

are still needed. Indeed, to achieve a more detailed modelling, various processes such as thermal processes, the ionization processes, and the field mapping in the gap should be considered. Also recall that the phenomena take place in three dimensions. Indeed, experimental and theoretical studies have emphasized the importance of the electric field in the initiation and development of electric discharge. It is well known that the slightest modification of space charge distributions may lead to local field enhancement in the air gaps and thereby affect discharge progression. As well, the initiation of discharge indicates a disturbance of the field. The dynamic distribution of the electric field within the gap might, therefore, be useful.

7.2 A general description of the dynamic procedure

The model is a self-consistent time-dependent one. The input data are the geometry of the electrodes, the voltage waveform, and the ambient air pressure and temperature.

The discharge time-base is divided into steps dt, starting from $t = 0$. At each time step, the applied voltage $U_c(t)$ is calculated. The inception conditions for the first corona (minimum threshold voltage and statistical time lag) are tested until they are satisfied, at which point formation of the first corona occurs. After the formation of the first corona, the leader inception conditions (the electric field at its tip) are tested until they are satisfied. Then the development of the second corona begins. Following the formation of the second corona, the propagation of the continuous streamer–leader system is initiated. The internal conditions of the leader channel (radius, gas number density, electric field, electrical parameters, and so on) and the unbridged gap capacitance are calculated.

The electric field at the leader tip is re-calculated and, if it exceeds a certain critical value, the propagation characteristics of the advancing system leader–streamer are determined, together with the leader velocity (determined by energy considerations) and current.

At each time step, the critical conditions for continued propagation of the discharge are tested and, if they are satisfied, the discharge continues to progress up to the beginning of the final jump. Otherwise, the leader–streamer system stops moving until new inception conditions are satisfied or stops definitively after the voltage crest.

During the final jump, the internal behaviour of the leader changes to LTE, which had not been the case before. The start of this phase takes place when the axial length of the leader reaches a certain value determined by the empirical relation defined by Garcia and Hutzler [3]. Finally, the arc phase follows inevitably unless the applied voltage drops significantly.

In the following sections, the different stages of the discharge are given with details of the main components of the model.

7.3 The applied voltage wave shape

7.3.1 The generation of a bi-exponential impulse

The voltage surges that stress the air gap clearances of power networks are of different forms; this concerns not only their amplitude but also their temporal variation. To study breakdown in large air gaps, HVs must be generated; the larger the gap distance, the higher the voltage to be applied. A Marx generator is a clever way of charging a number of HV capacitors in parallel, then discharging them in series via spark gaps into 'shaping' circuits of resistors and capacitors. Originally developed by E Marx (1924), Marx generators are probably the most common way of generating HV surges for testing when the voltage level required is higher than available charging supply voltages. The output voltages, the so-called impulse voltages, are used to simulate internal surge voltage (switching surge) and external surge voltage (lightning surge) along with their consequences for different dielectrics of gaseous, liquid, or solid origins. Measurement may be accomplished by means of a capacitive and/or resistive voltage divider and a fast digital oscilloscope with storage facilities. Impulse voltages for HV testing are standardized according to IEC 60-2.

The most important basic circuits are called type a (low efficiency), and type b (high efficiency), respectively (figure 7.1).

In the figure, C_g is the discharge capacitor; G the spark gap; R_c the damping resistor (for shaping the front); R_g the discharge resistor (for shaping the tail); and C_c is the load capacitor (the capacitance of the equipment under test). Marx generators are characterized by the maximum output voltage and the maximum storable electrostatic energy. The maximum discharge energy can be computed from the following equation:

$$W = \tfrac{1}{2} C_g U_0^2. \tag{7.1}$$

The capacitance (C_g) is charged slowly (charging time constant between 10–20 s) up to the desired voltage (U_0). When fully charged, either the spark gap G is allowed to breakdown from over-voltage or it is triggered by an external source (if the gap spacing is set greater than the charging voltage breakdown spacing). Since R_c, the damping resistor (for shaping the front), is much lower than R_g, the discharge resistor (for shaping the tail), this effectively puts the capacitors C_g and C_c in series. Capacitor C_c (and the eventual residual charges of capacitor C_g) then discharges in the resistances R_c and R_g. More detail concerning Marx generators and multi-stages theories can be found in HV engineering textbooks. The basic principles are recalled hereafter.

The normalized wave shapes for lightning and switching surges are presented in figure 7.2.

The normalized (standard) wave shape for the lightning impulse voltage is:

1.2/50 μs, $T_1 = 1.2$ μs ± 30%, and $T_2 = 50$ μs ± 20%.

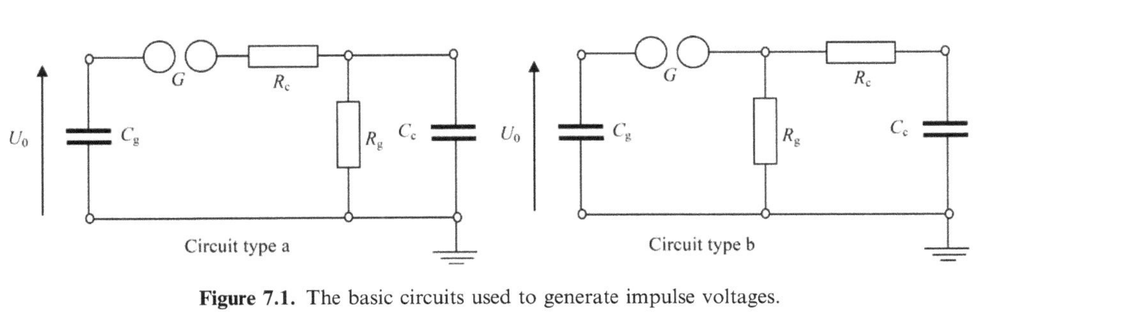

Figure 7.1. The basic circuits used to generate impulse voltages.

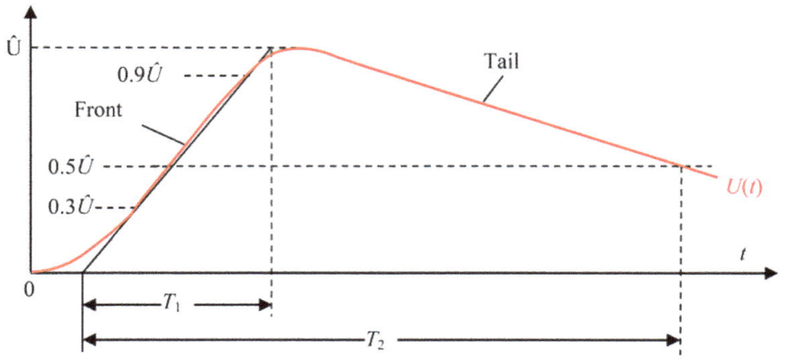

Figure 7.2. Normalized impulse voltages.

The normalized (standard) wave shape for the switching impulse voltage is:

250/2500 μs, $T_1 = 250$ μs ± 20%, and $T_2 = 2500$ μs ± 60%.

The maximum content of high frequency oscillations is $< 0.05\hat{U}$ with $f > 0.5$ MHz. When $f < 0.5$ MHz one has to consider $\hat{U}$ = actual highest voltage peak.

The wave shape of these surges is characterized by two parameters (T_1 and T_2):

- the wave front $T_1 = 1.67(T_{90}-T_{30})$ and
- the wave tail times T_2 is the time to mid-amplitude.

Each impulse voltage is therefore represented by T_1/T_2, both parameters (T_1 and T_2) being expressed in μs.

The voltage wave shape considered for the mathematical model is of type b because most Marx generators used in HV laboratories are of this type. They are common due to their higher possible utilization factor.

The voltages, applied to air gaps, were obtained using a single-stage impulse generator circuit (figure 7.1(b)). Considering Laplace transformation, the output voltage can be derived [13]:

$$\frac{U_0}{p} = I_1 R_g + (I_1 + I_2)\frac{1}{pC_g} \tag{7.2}$$

$$I_1 R_g = I_2\left(R_c + \frac{1}{pC_c}\right) \tag{7.3}$$

$$U_c(p) = \frac{I_2}{pC_c}. \tag{7.4}$$

Substituting equations (7.2) and (7.3) into (7.4) yields, after rearrangement,

$$U_c(p) = \frac{U_0}{R_c C_c}\frac{1}{p^2 + bp + c} \tag{7.5}$$

with

$$b = \frac{1}{R_c C_c} + \frac{1}{R_g C_g} + \frac{1}{R_c C_g} \tag{7.6}$$

$$c = \frac{1}{R_c R_g C_c C_g}. \tag{7.7}$$

Equation (7.5) may be rewritten as

$$U_c(p) = \frac{U_0}{R_c C_c} \frac{1}{\left(p + \frac{1}{T_a}\right)\left(p + \frac{1}{T_b}\right)}, \tag{7.8}$$

where $1/T_a$ and $1/T_b$ are the roots of the equation

$$p^2 + bp + c = 0 \tag{7.9}$$

The parameters T_a and T_b may then be derived

$$T_{a,b} = \frac{2}{b \pm \sqrt{b^2 - 4c}}. \tag{7.10}$$

The inverse Laplace transform of equation (7.8) provides the temporal evolution of the output voltage

$$U_c(t) = \frac{U_0}{R_c C_c} \frac{T_a T_b}{T_a - T_b} \left[\exp\left(-\frac{t}{T_a}\right) - \exp\left(-\frac{t}{T_b}\right) \right], \tag{7.11}$$

where T_a, T_b, T_c and U_0 are the rise time, the fall time, the time to peak value and the amplitude factor, respectively.

The time constants T_a and T_b are found in the literature [13],

$$T_a = \frac{T_2 - T_{cr}}{0.7} \tag{7.12}$$

and

$$T_b = \frac{T_a}{\exp\left(1.35 + 1.2 \ln \frac{T_2}{T_{cr}}\right)}. \tag{7.13}$$

The maximum value of $U_c(t)$ will be for

$$t = T_{cr} = \frac{T_a T_b}{T_a - T_b} \ln \frac{T_a}{T_b} \tag{7.14}$$

with

$$\frac{T_a T_b}{(T_a - T_b) R_c C_c} = 1. \tag{7.15}$$

The output voltage $U_c(t)$ can be rewritten as

$$U_c(t) = U_0\left(\exp\left(-\frac{t}{T_a}\right) - \exp\left(-\frac{t}{T_b}\right)\right). \tag{7.16}$$

From equation (7.8), time constants T_a and T_b, can be deduced

$$T_a \approx R_c\left(C_g + C_c\right) \tag{7.17}$$

$$T_b \approx R_g\frac{C_gC_c}{C_g + C_c}. \tag{7.18}$$

Considering $\delta = \frac{T_a}{T_b}$, U_0 can be deduced as function of the crest voltage U_{cr} [14],

$$U_0 = \frac{U_{cr}}{\left(\delta^{1/(1-\delta)} - \delta^{\delta/(1-\delta)}\right)}. \tag{7.19}$$

The capacitance C_c can be derived from equation (7.20)

$$C_c = \frac{T_aT_b}{T_a - T_b}\frac{1}{R_c}. \tag{7.20}$$

The flowchart given in figure 7.3 summarizes the procedure used to determine the electrical parameters of figure 7.1(b), once the expected voltage output $U_c(t)$ is characterized.

7.3.2 Superimposed oscillations

One may superimpose voltage oscillations on the bi-exponential impulse using an inductance L_m and capacitance C_m which are fitted between the capacitive divider and the HV electrode. The time development of the voltage can be derived using the equivalent circuit shown in figure 7.4, in which the frequency of oscillation is given by:

$$f = \frac{1}{2\pi}\sqrt{\frac{1}{L_mC} - \left(\frac{R}{2L_m}\right)^2} \tag{7.21}$$

where R and C are

$$R = \sqrt{2}\frac{L_m}{\left(R_g + R_c\right)\left(\frac{C_c}{C_m}\right)(C_m + C_c)} \tag{7.22}$$

$$C = \frac{C_mC_c}{C_m + C_c}. \tag{7.23}$$

The condition for aperiodicity is: $R^2 > 4L/C$ for a dc voltage wave.

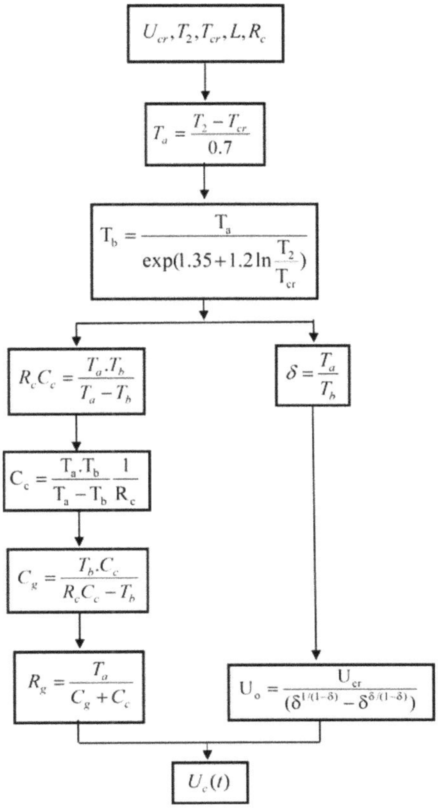

Figure 7.3. Block diagram for the impulse voltage calculation.

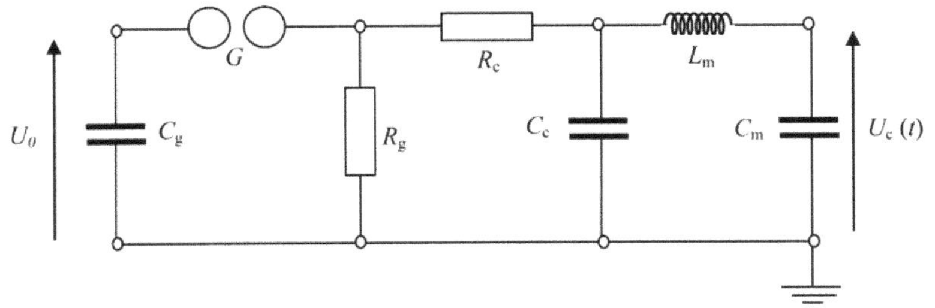

Figure 7.4. Basic circuits used to generate oscillating impulse voltages.

7.4 The characterization of the discharge propagation

7.4.1 The inception of the first corona

From shot to shot, with the same experimental conditions, the inception time and voltage can vary. This is simulated by a random time lag dt_0, following the theoretical inception time, due to the need for a free electron near the HV electrode. As a consequence, the inception time is the sum of the theoretical inception time and dt_0, which has a uniform probability distribution in the model. The delay means that the over-voltage at inception is higher than its theoretical value U_{0i}.

The theoretical inception voltage for the first corona, $U_{0i,}$ is deduced from Peek's law for the point–plane geometry [15] to become

$$U_{0i} = \frac{1}{2} R_p E_0 \log\left(\frac{4D}{R_p}\right)\left(1 + \frac{0.0436}{\sqrt{R_p}}\right), \tag{7.24}$$

R_p being the HV electrode's radius, D the gap length, and E_0 the critical field at which the ionization phenomena become significant. It is known that the surface field can be affected by surface conditions, such that, according to Peek's experimental law, $E_0 = 31\delta$ (in kV cm^{-1}). Furthermore, δ is the relative air density factor [16]

$$\delta = \frac{0.392b}{273 + \alpha} = \frac{2.92c}{273 + \alpha}, \tag{7.25}$$

where b is the barometric pressure in mmHg, c is the barometric pressure in kPa, and α is the temperature in kelvins.

The length of the first corona-streamers is obtained using the following empirical expression [15]:

$$S_0 = 1.44 U_{0i} \quad (MV, \text{m}), \tag{7.26}$$

with U_{0i} in megavolts and S_0 in metres. The development of the leader is disturbed by the significant space charge left by the first corona. The inception voltage, U_i, has an influence on the duration of the discharge until the final jump [1].

7.4.2 The criterion for advancement of the leader tip

The field E_p at the leader tip is an important parameter governing its growth. The idea of the model consists in replacing the positive leader tip with an equivalent electrode [7]. If the electric field at this electrode surface is sufficient (higher than a critical field E_c) to keep the corona effect alive, the leader channel can propagate through the gap. Otherwise, the leader channel progressively cools and finally disappears, unless the applied voltage characteristics are able to re-activate it. The field E_p can be estimated using a hyperboloid approximation [7], namely:

$$E_p = \frac{2U_j}{R_e \ln 4\left(\frac{D - L_Z}{R_e}\right)}, \tag{7.27}$$

where U_j is the voltage at the leader tip and R_e is the radius of the equivalent hyperboloid electrode. An empirical expression for R_e was proposed by [7], namely:

$$R_e = \frac{U_j}{U_{j0}} \frac{0.25}{1 + \frac{3.5}{L}} + R_i \exp\left(\frac{L_Z}{-13.6R_i}\right), \tag{7.28}$$

where R_i, L, and L_Z are the initial equivalent radius after the development of the first corona, and the leader's effective and axial lengths, respectively, while U_{j0} is the voltage to be applied to the unbridged gap to render the field on the equivalent electrode surface equal to the critical field E_c [7], given by:

$$U_{j0} = \frac{Ec_0 R_x}{2} \ln 4\left(\frac{D - L_Z}{R_x}\right), \tag{7.29}$$

where

$$Ec_0 = E_0\left(1 + \frac{0.0436}{\sqrt{\delta R_x}}\right), \tag{7.30}$$

$$R_x = \frac{0.25}{1 + \dfrac{3.5}{L}} + R_i \exp\left(\frac{L_Z}{-13.6R_i}\right). \tag{7.31}$$

The field E_c, depends on the electrode shape and varies in the range 30–40 kV cm^{-1} for typical electrodes. For a long time, Peek's law, described in chapter 3, has been used in overhead network problems. A generalization of this law [6] has been proposed for the case of electrodes with spherical symmetry. The field E_c at a given pressure p, is given by

$$E_c = E_0\frac{p_0}{p}\left\{1 + \left(\frac{\left(\frac{p_0}{p}\right)^2 - 1}{2}\right) + \left[\left(\frac{\left(\frac{p_0}{p}\right)^2 - 1}{2}\right)^2 + 9.510^{-4}\frac{p_0}{p}\left(\frac{1}{R_1} + \frac{1}{R_2}\right)\right]^{\frac{1}{2}}\right\}, \tag{7.32}$$

where R_1, R_2, and p_0 are the radii of curvature and atmospheric pressure, respectively. At atmospheric pressure, in the case of an electrode with a spherical extremity ($1/R_1 = 1/R_2 = 1/R_e$), it yields:

$$E_c = E_0\left[1 + \frac{0.0436}{\sqrt{R_e\delta}}\right]. \tag{7.33}$$

Instead of field calculations using classical numerical methods, such as equivalent charges, finite elements and the Monte Carlo method which all require long computations, the above analytical relationship can be used to describe the experimental results fittingly. A random statistical time lag dt_0 is also considered. The simplification owing to the hyperbolic approximation has been described in [7]. Note that such a simplification is limited to the case of a plane cathode.

7.4.3 The propagation of a leader corona

Leader propagation may be represented by successive straight segments of length dL ($dL = v dt$, while v and $dt = 10^{-6}$ s are the leader propagation velocity and the digital sample, respectively) which make a random angle with the gap axis, such as $dL_Z = dL\cos$; dL_Z is the length of the leader step axial.

In [17], the authors extended the formula of velocity established by Beroual [18, 19] for discharges propagating in liquid dielectrics to predict the leader propagation velocity in long air gaps. At any time, the leader propagation velocity is computed, provided the energy (or power) injected into the gap (taking into account the specific conditions, such as the voltage and the gap geometry) is known

$$v_j(t) = \left(\frac{2\beta}{\rho \pi r_0^2} \frac{\mathrm{d}W_j(t)}{\mathrm{d}t} \right)^{1/3} \tag{7.34}$$

or

$$v_j(t) = \left(\frac{2\beta}{\rho \pi r_0^2} P_j(t) \right)^{1/3}, \tag{7.35}$$

with $\rho = 0.1$–$0.15\ \mathrm{kg\ m^{-3}}$ and r_0 the leader radius at the time t_0 when it is formed. β is the fraction of available energy used for the displacement of discharge; a typical value was estimated so that $\beta = 0.1$ [17].

From digital samples of the instantaneous voltage U_j, and the HV current I_j, the power P_j, and energy W_j injected into the gap are obtained from successive computations:

$$P_j = U_j I_j \tag{7.36}$$

$$W_j = \sum_0^t U_j I_j \mathrm{d}t. \tag{7.37}$$

Figure 7.5 represents a simplified sketch of the discharge geometry for a rod–plane gap of length D. Sub-dividing the voltage drop across the gap into three terms (the leader, streamer, and un-bridged gap terms), the following can be obtained, for each step,

$$U_c(t) = E_L L_Z + E_S S + E_Z Z, \tag{7.38}$$

where E_L, E_S (roughly constant and equal to 0.45–0.5 MV $\mathrm{m^{-1}}$ [1, 2]), and E_Z are the mean fields and L_Z, S, and Z the axial lengths for the three discharge regions (see figure 7.5). E_L can be calculated using non-LTE considerations [2] with an initial value at 1.5 kV $\mathrm{cm^{-1}}$,

$$E_L = \frac{E_L}{N} \frac{N_0}{\sqrt{1 + 2\frac{\gamma - 1}{\gamma} \frac{E_L}{N} \frac{Q(t)}{k \tau_0 \pi r_0^2}}} \tag{7.39}$$

where N ($N = \sigma/\mu_e e \xi$) is the gas number density in the channel core of radius r. τ_0, N_0, are the gas temperature and the gas number density, respectively, at time t_0 when the leader section is formed. Further, k is the Boltzmann constant ($k = 1.38 \times 10^{-23}\ \mathrm{J\ K^{-1}}$), μ_e the electron mobility ($\mu_e = 0.18\ \mathrm{m^2\ V^{-1} s^{-1}}$), e the elementary electron charge ($e = 1.6 \times 10^{-19}$ C), and ξ the ionization degree. The ratio γ, is taken between specific heats at constant volume and constant pressure, by

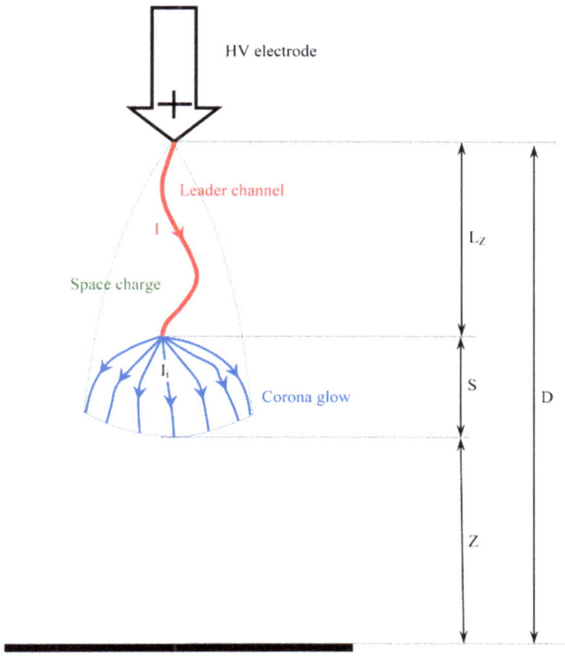

Figure 7.5. Schematic representation of the leader–streamer system. Reproduced from [20] with permission.

hypothesis, constant and equal to 1.4 [2]. The ratio E_L/N, called the reduced field, along with the ionization degree and the channel conductivity are related to the current I flowing in the leader channel. Their characteristics have been established by Les Renardières [1] (figure 7.6).

$Q(t)$ is defined as

$$Q(t) = \int_{t_0}^{t} I \mathrm{d}t. \tag{7.40}$$

The main problem is to obtain a good estimate for E_Z. As the known gap distance is $D = L_Z + S + Z$, the following expression was proposed for E_Z:

$$E_Z = \frac{U_c(t) - U_i}{D}. \tag{7.41}$$

The length of the leader corona-streamers can be deduced from equation (7.38) as:

$$S = \frac{U_c(t) - E_Z Z - E_L L_Z}{E_S}. \tag{7.42}$$

The streamer front advancement velocity is deduced at each time as $v_s(t) = \mathrm{d}S/\mathrm{d}t$. Under specific circumstances the streamer velocity may go to zero, arresting the current that is feeding the leader tip. The fulfilment of the condition $v_s(t) > 0$ is therefore checked at every time.

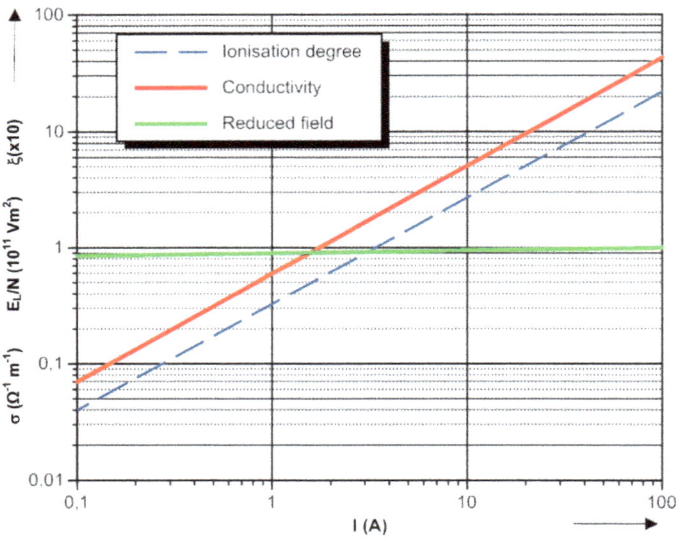

Figure 7.6. Characteristics of the leader channel as a function of current with $r_0 = 1$ mm and $T_0 = 1000$ K [1]. Figure reprinted with permission from CIGRE, Positive discharges in long air gaps © 1977.

7.4.4 Dispersion

It is known that the hazardous propagation of the leader and dispersion on the inception voltage is responsible for the dispersion of voltages and at times leads to flashover. Hutzler and Hutzler [7] have established a probabilistic law, which seems to be physically plausible. The erratic propagation is taken into account with a law of probability characterized by a dispersion parameter a. The probabilistic density $P(\varphi)$ of the angle is

$$P(\varphi) = \frac{\sin\varphi}{a}\exp\left(-\frac{1-\cos\varphi}{a}\right). \tag{7.43}$$

It is possible to explain the parameter a in the following form [7]

$$a = 0.6(1+10R_t)^{-1}\left(0.8 + 10^{-10}\frac{\mathrm{d}U_c(t)}{\mathrm{d}t}\right)^{-1} \text{(m, V, s)}, \tag{7.44}$$

where R_t is the curvature radius of the equipotent surface at the considered point and $\mathrm{d}U_c(t)/\mathrm{d}t$ (in V s^{-1}) is the rate of variation of the applied voltage at the considered time. This empirical formula expresses the fact that the erratic behaviour of the leader is more marked when the field is less uniform and when voltage increases less rapidly.

From equation (7.43) it may be observed that the probability of having $\varphi > \pi/2$ is non-null; the possibility for the leader to travel back towards the HV electrode is, therefore, allowed in the model. Since it has no physical significance, it is necessary to truncate equation (7.42):

$$P(\varphi) = \frac{|\sin\varphi|}{2a\left(1-\exp\left(\frac{-2}{a}\right)\right)}\exp\left(\frac{\cos\varphi - 1}{a}\right). \tag{7.45}$$

7.4.5 The final jump

The transition from the free development phase to the final jump phase will be achieved when the value of L_Z (computed at each time as $L_Z = L_Z + \mathrm{d}L_Z$ with an initial value at $L_Z = 10\ \mu\text{m}$) will be higher or equal to L_{fj}, the so-called final jump height, deduced from the empirical relation proposed by Garcia and Hutzler [3]

$$L_{\text{fj}} = D - \frac{U_{\text{c}}(t) - 0.15}{0.44}(\text{MV},\ \text{m}). \tag{7.46}$$

Generally, the initiation of this phase inevitably leads to breakdown, except when the applied voltage drops notably. If the mean streamer field E_{ms} [21]

$$E_{\text{ms}} = \frac{U(t) - 0.15}{D - L_Z}(\text{m},\ \text{MV}) \tag{7.47}$$

is high enough, i.e. higher than E_{s}, the leader channel can propagate at a high velocity of few metres per second [4]. Otherwise, the gap withstands the voltage surge.

7.4.6 The return stroke current

LTE applies in the channel, i.e. during the arc phase. This arc is a transient one which discharges the HV generator capacitors. The gap voltage drops entirely. Particle recombination is the dominant phenomenon. This phase is characterized by the flow of a high current in the discharge channel. The value of this current is affected by the characteristics of the test circuit. For gap lengths below 10 m, this current is generally lower than 10 kA [22, 23].

7.5 Distributed-circuit-based modelling

It is already known that circuit models have a valuable role to play in the understanding of the plasma and some related processes [24–28]. It is assumed that the discharge can be represented by an equivalent electrical network represented in figure 7.7, where R_0 is the plane electrode resistance and n is the number of leader steps. A value of $R_0 = 100\ \Omega$ is assumed [24, 26], while C_0 is taken as a half spherical capacitance, estimated from a spherical approximation (the gap geometry is assimilated to two concentric spheres, of radius R_{p} and $\{R_{\text{p}} + D\}$)

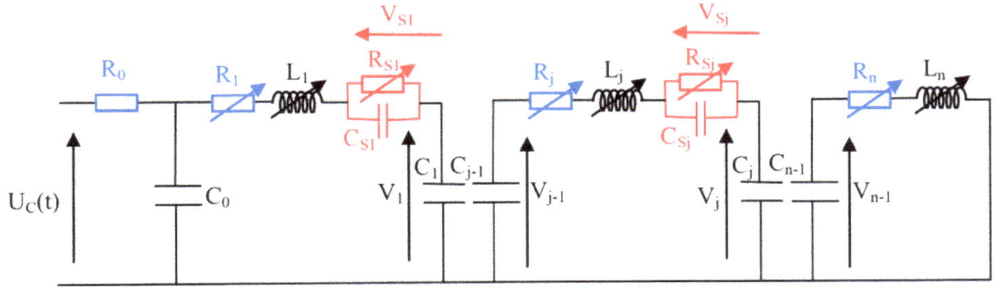

Figure 7.7. The distributed-circuit model of the discharge. Each value of j represents a different discharge step. Reproduced from [20] with permission.

$$C_0 = 2\pi\varepsilon_0 \frac{(D + R_p)R_p}{D}. \tag{7.48}$$

When the leader propagates, the initiated potential and current can be computed as follows:

$$L_j \frac{\mathrm{d}i_j(t)}{\mathrm{d}t} = V_{j-1}(t) - V_j(t) - R_j i_j(t) - Vs_j(t) \tag{7.49}$$

$$C_{sj}(x) \frac{\mathrm{d}Vs_j(t)}{\mathrm{d}t} = i_j(t) - \frac{Vs_j(t)}{Rs_j} \tag{7.50}$$

$$C_j(x) \frac{\mathrm{d}V_j(t)}{\mathrm{d}t} = i_j(t). \tag{7.51}$$

On the other hand, the capacitive currents are

$$i_{j-1}(t) - i_j(t) = C_{j-1}(x) \frac{\mathrm{d}V_{j-1}(t)}{\mathrm{d}t}. \tag{7.52}$$

To ease computation of these equations, the 'state representation' was considered. This consists in writing the circuit equations in a system of first order differential equations

$$\frac{\mathrm{d}X(t)}{\mathrm{d}t} = AX(t) + BJ(t) \tag{7.53}$$

$$Y(t) = GX(t) + HJ(t), \tag{7.54}$$

where A, B, G, and H are the state matrices corresponding, respectively, to the dynamic of the system, the control, the observation, and the direct transmission.

The initial matrices are:

$$A = \begin{bmatrix} -(R_g + R_c)/(C_g R_c R_g) & 1/(R_c C_g) & 0 \\ 1/(R_c C_c) & -(R_0 + R_c)/(C_c R_c R_o) & 1/(R_0 C_c) \\ 0 & 1/(R_0 C_0) & -1/(R_0 C_0) \end{bmatrix}; \; B = \begin{bmatrix} 0 \\ 0 \\ 0 \end{bmatrix};$$

$G = B^T$ and $H = [0]$ for a pure impulse voltage or:

$$A = \begin{bmatrix} -(R_g + R_c)/(C_g R_c R_g) & 1/(R_c C_g) & 0 & 0 & 0 \\ 1/(R_c C_c) & -1/(R_c C_c) & 0 & -1/C_c & 0 \\ 0 & 0 & -1/(R_0 C_m) & 1/C_m & 1/(R_0 C_m) \\ 0 & 1/L_m & -1/L_m & 0 & 0 \\ 0 & 0 & 1/(R_0 C_0) & 0 & -1/(R_0 C_0) \end{bmatrix}; \; B = \begin{bmatrix} 0 \\ 0 \\ 0 \\ 0 \\ 0 \end{bmatrix};$$

$G = B^T$ and $H = [0]$ for a superimposed oscillating voltage. $J(t)$ and $Y(t)$ are the input and output matrices, respectively.

Once the state vector X is determined, the state matrices (A, B, G, H) are computed knowing the topology and the values of the circuit parameters. Substituting equations (7.49), (7.50) and (7.51) into (7.52), voltage drops are computed. Then, the leader current is deduced

$$i(t) = \frac{U_c(t) - V_0(t)}{R_0}, \tag{7.55}$$

$V_0(t)$ and $U_c(t)$ being the capacitance C_0 voltages. An initial value of 0.1 mA is considered at the start up. To guarantee convergence, the simulated value of $i(t)$ should be kept higher than 30 mA. This value has been derived from experimental recordings [1, 6, 9, 21].

The space charge injected into the gap is computed as the sum of the elementary charges, i.e.

$$Q(t) = \sum_1^n q_j = \sum_0^n \sum_0^t C_j V_j(t), \tag{7.56}$$

$V_j(t)$ being the C_j capacitance voltage, while n is the number of leader steps.

Once the elements L_j, C_j, and R_j are known, together with the applied voltage wave, the spatial and temporal discharge characteristics (current, charge, etc) can be computed.

The number of cells is increased each time the field E_p at the leader tip is higher than the critical field. The state space matrices are then recalculated:

for $j = 1 : n$,

$$A(2{*}j + 1 + ao,\ 2{*}j + 2 + ao) = -1/C(j);$$
$$A(2{*}j + 2 + ao,\ 2{*}j + 2 + ao) = -R(j)/L(j);$$
$$A(2{*}j + 2 + ao,\ 2{*}j + 1 + ao) = -1/L(j);$$
$$A(2{*}j + 3 + ao,\ 2{*}j + 2 + ao) = 1/C(j + 1);$$
$$A(2{*}j + 2 + ao,\ 2{*}j + 3 + ao) = -1/L(j);$$

end;

$B = \text{zeros}(2{*}n + b2, 1)$; $G = B^T$ and $H = [0]$ with ($ao = 0$; $b1 = 2$ and $b2 = 3$ for a pure impulse voltage) and (ao = 2; b1 = 3 and b2 = 5 for a superimposed oscillating voltage).

7.6 The distributed-circuit elements

7.6.1 The resistance

The leader corona

The leader corona consists of many filamentary channels or streamers as shown in figure 7.5. Let us call I_i the current flowing in each branch, so $\Sigma I_i = \text{I}$. Because of the high value of the streamer conductivity, I_i is very small with respect to I. Recall from chapter 3 that electrons flowing at the tip of the leader induce a concomitant heating of the common root of the streamers. This zone, known as the stem, represents the transition from the streamer to the leader. The translation of this process into a mathematical model was eased by making a simplified approximation, considering the leader corona of length S as a part of the leader.

The leader channel

Leader propagation phase
During the leader propagation, no LTE applies within the channel. This channel serves as a resistive electrode, the properties of which are controlled by Ohmic heating and gas dynamic expansion [1, 4]. Assuming the channel to be a cylindrical one, the resistance per unit length can be computed as follows:

$$\lambda = \frac{1}{\sigma \pi r^2 (t - t_0)}. \tag{7.57}$$

The leader conductivity σ can be deduced from the current I [1, 2]. Figure 7.6 summarizes the characteristic $\sigma = f(I)$ based on works performed by the Les Renardières group [1]. This graph was obtained considering initial conditions at $r_0 = 1$ mm and $T_0 = 1000$ K. Since the value of r_0 depends on the gap length, this characteristic needs to be truncated. For a given gap length with $r_{0\mathrm{r}}$ as the initial radius of the leader channel, the actual conductivity σ_r is adjusted as

$$\sigma_\mathrm{r} = \left(\frac{r_0}{r_{0\mathrm{r}}}\right)^2 \sigma = ct^2\sigma. \tag{7.58}$$

The velocity (equation (7.34) or (7.35)) should also be multiplied by ct. Under non-LTE conditions, the cross-section of the leader channel, at any time t and position x, is directly related to the charge flow [2]:

$$(\pi r^2)_{x,t} = \sqrt{(\pi r_0{}^2)^2 + 2\frac{\gamma - 1}{\gamma}\frac{E_L}{N}\frac{\pi r_0{}^2}{k\tau_0}Q(t)}\,. \tag{7.59}$$

Typical values of r_0, obtained from experimental investigations [4, 21, 29], are represented in figure 7.8 as a function of the gap length.

The final jump phase
The final jump represents a transition during which the conditions within the leader channel change from non-LTE to LTE characteristics of the spark channel. The internal conditions change in a much shorter time, almost constant and equal to 20 μs, regardless of the gap geometry and applied voltage wave form [30]. The increase in the degree of ionization is accompanied by the concomitant reduction in the per unit length resistance of the leader. From experimental data reported by Gallimberti [4], an empirical relationship was proposed (regression) to model the resistance variation R_j during this phase:

$$R_j = 10^{\sqrt{p^2 - \frac{t}{\Delta t}}}, \tag{7.60}$$

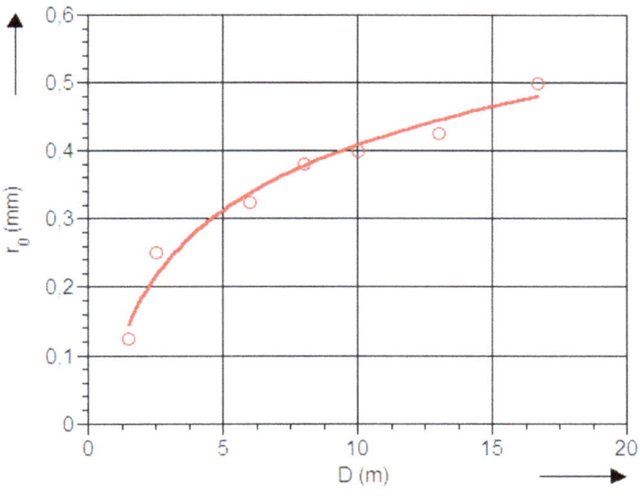

Figure 7.8. The variation of the leader's initial radius versus the gap length extrapolated from experimental results. Figure reproduced from [29] with permission of Université de Pau et des Pays de l'Adour, France.

where Δt represents the duration of the final jump phase and $p = log_{10}R_{fj}$; R_{fj} being the last value of the leader resistance R_j just before the beginning of the final jump.

The return stroke

During this phase, LTE applies within the highly ionized channel [4]: the gas molecules are fully dissociated, and the temperature is uniform throughout the channel. This phase corresponds to the neutralization of charges injected into the gap during the leader–streamer propagation. The channel radius is maintained by magnetic and kinetic forces [27]. The conductivity of this highly ionized gas is given by $\sigma = 1.5 \times 10^{-5}\tau^{3/2}(\Omega^{-1}\mathrm{cm}^{-1})$. During this phase, the temperature within the channel may reach values as high as 20 000 K for long air gaps [4]. The channel is simply a low impedance path to the ground.

7.6.2 The inductance

Considering the leader channel as a current-carrying line, its inductance may be determined. The simplified assumptions proposed in the literature [26, 27] are considered. The end effects are neglected. The per unit length inductance L_u of the channel is the sum of two inductances [26], namely:

- the inductance due to internal flux (internal inductance)

$$L_1 = \frac{1}{4}\frac{\mu_0}{2\pi}(\mathrm{Hm}^{-1}); \tag{7.61}$$

- the inductance due to external flux (external inductance)

$$L_2 = \frac{\mu_0}{2\pi}\ln\left(\frac{D_f}{r}\right)(\mathrm{Hm}^{-1}); \tag{7.62}$$

where r represents the leader channel radius, and D_f is the distance from the channel where the magnetic field drops to zero.

The total inductance ($L_1 + L_2$) reads

$$L_u = \frac{\mu_0}{2\pi}\left(0.25 + \ln\left(\frac{D_f}{r}\right)\right)(\text{H m}^{-1}); \tag{7.63}$$

Note that D_f should be large for transient fields and that the fractional error is of the order of $1/\ln(D_f/r)$ [27]. This error is very small since $D_f > r$. For laboratory long air gap discharges, a value of $D_f = 100$ m was acceptable [26]. The duration of the leader inception and propagation phase were much larger than that of the final jump and return stroke, the effect of the inductances can be neglected during leader phase.

7.6.3 The capacitance

The not-bridged gap

The capacitance C_j between the front head of the corona-streamers and the grounded plane electrode can be computed by spherical approximation [24–26]. In this simplified approximation, the leader–streamer system is represented as a plasma channel surrounded by an envelope of positive charges induced by the foregoing streamers. This approach consists of assuming the discharge head and the plane to be two concentric spheres. The distance between both spheres varies with leader–streamer advancement (figure 7.9).

Since the solid angle Γ ($\Gamma = 2\pi(1 - \cos\theta)$) is lower than 4π, the capacitance can be deduced, as follows:

$$C_j = \Gamma_j \varepsilon_0 \frac{D(L_Z + S)}{Z}. \tag{7.64}$$

During each leader step, the elementary charge injected can be expressed as

$$q_j = C_j U_j = \Gamma_j \varepsilon_0 \left[\frac{D(L_Z + S)}{Z}\right] U_j. \tag{7.65}$$

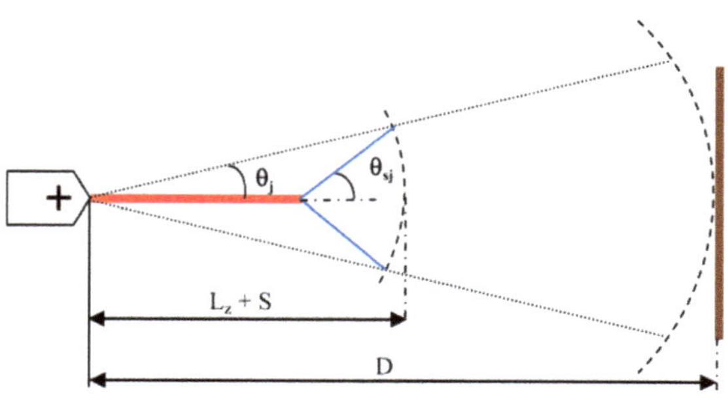

Figure 7.9. Spherical approximation considered in the model. Reproduced from [20] with permission.

The corresponding current flowing in the channel can be deduced from the derivate of the elementary charge (equation 7.65)

$$I = \frac{\mathrm{d}}{\mathrm{d}t}\left[\Gamma_j\varepsilon_0\frac{D(L_Z + S)}{Z}U_j\right]. \tag{7.66}$$

The differentiation of this equation allows the deduction the leader current

$$I = \frac{\mathrm{d}\Gamma_j}{\mathrm{d}t}\varepsilon_0\left(\frac{D(L_Z + S)}{Z}\right)U_j + \frac{\mathrm{d}U_j}{\mathrm{d}t}\Gamma_j\varepsilon_0\left(\frac{D(L_Z + S)}{Z}\right) + \Gamma_j\varepsilon_0U_j\frac{\mathrm{d}}{\mathrm{d}t}\left(\frac{D(L_Z + S)}{Z}\right). \tag{7.67}$$

The solid angle Γ_j may not remain constant during the leader–streamer advancement in the gap (figure 3.12). Let us assume a variation $\mathrm{d}\Gamma_j = \Gamma_j - \Gamma_{j-1}$ between steps, knowing that $\mathrm{d}S/\mathrm{d}t = v_s$ and $(\mathrm{d}L_Z/\mathrm{d}t)\cos\varphi = v\cos\varphi$, the variation of Γ_j can be derived as

$$\Gamma_j = \frac{\dfrac{I}{\varepsilon_0} + \dfrac{D(L_Z + S)}{Z\mathrm{d}t}U_j\Gamma_{j-1}}{\dfrac{D(L_Z + S)}{Z\mathrm{d}t}U_j + \dfrac{\mathrm{d}U_j}{\mathrm{d}t}\dfrac{D(L_Z + S)}{Z} + U_j\left(\dfrac{M - N}{Z^2}\right)}, \tag{7.68}$$

where

$$\begin{cases} M = (D^2 - D(L_Z + S))(v_L\cos\varphi + v_S) \\ N = -(v_S + v_L\cos\varphi)(D(L_Z + S)) \end{cases}. \tag{7.69}$$

The leader corona

The capacitance C_{sj} between the leader tip and the front head of the corona–streamers can also be computed by the spherical approximation concept. The radius of the concentric sphere to be considered in this case is the radius of the leader tip (r_0) and the radius of the space-charge delimited front head of the corona-streamers (S). Considering that the solid angle θ_{sj} ($\Gamma_{sj} = 2\pi(1 - \cos\theta_{sj})$) is lower than four, and S_{r0}, C_{sj} can be estimated as

$$C_{sj} = \Gamma_{sj}\varepsilon_0\left(\frac{(S + r_0)r_0}{S}\right) = \Gamma_{sj}\varepsilon_0 r_0. \tag{7.70}$$

7.7 General flowchart of the model

The time constant R_{sj}, C_{sj} is very small with respect to R_jC_j. Therefore C_{sj} may be neglected. Camera recordings (in frame mode) of discharge propagating in long air gaps reveal that leader corona can be seen as a conical luminous region with an angle of about 70° [1], whose apex is near the leader tip. To start the simulation a conical angle of the first corona, $\theta_0 = \pi/6$, is considered.

The block diagram given in figure 7.10 describes the different steps of the model, where T_M represents the maximum duration scheduled for the total discharge development for a given configuration (gap geometry, applied voltage and atmospheric conditions). This block diagram presents several loops. The equilibrium values corresponding to stable propagation are imposed by the applied voltage, which represents the external input to the system. Statistical fluctuations or external disturbances may generate oscillations in the different loops.

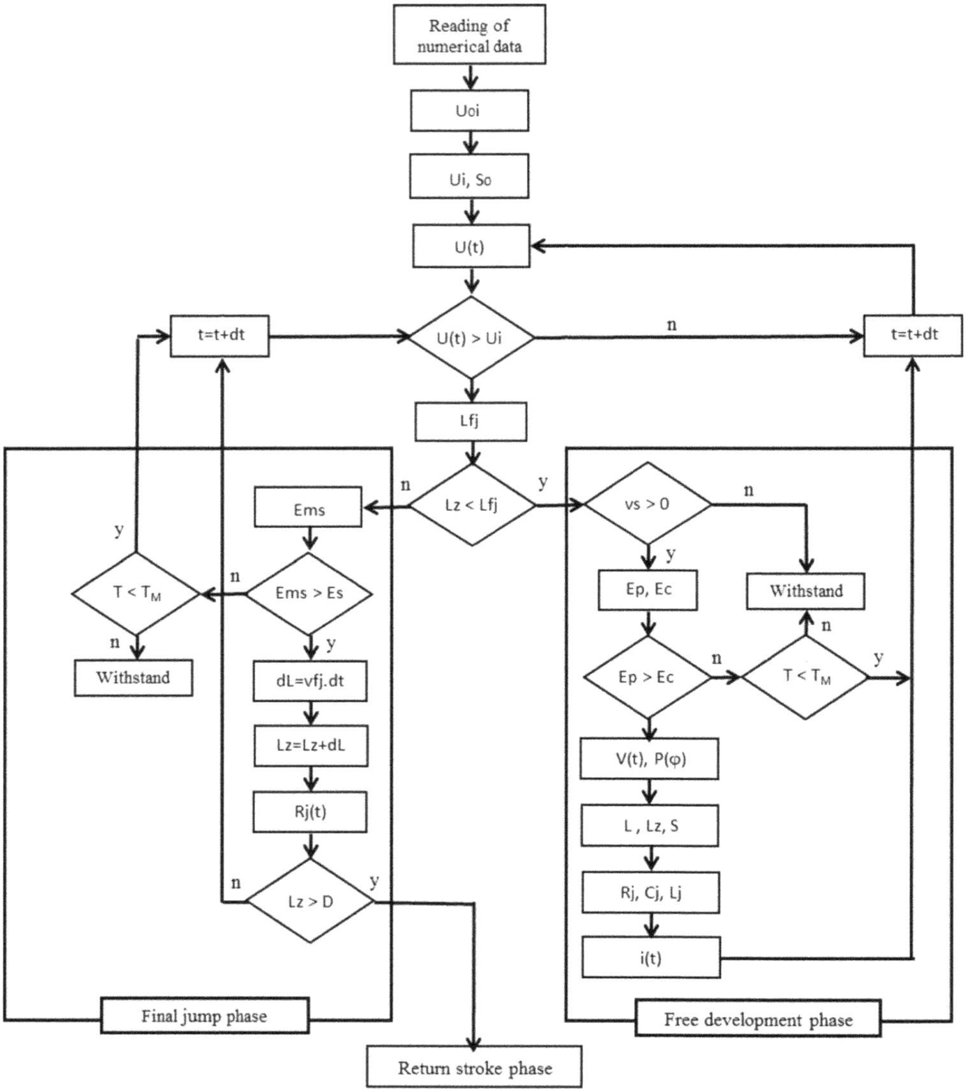

Figure 7.10. Flowchart of the discharge propagation. Reproduced from [20] with permission.

7.8 Extension to a very long air gap: positive lightning

Lightning can be seen as the longest natural spark discharge. From a visual point of view, lightning and a laboratory spark look similar, differing only in scale. A mathematical model of the positive lightning downward leader was developed based on the idea that a substantial similarity exists between lightning phenomena and the discharge in large air gaps. The same concept was, therefore, used to simulate lightning discharge. In this section, important changes with regard to the model developed and depicted by the flowchart of figure 6.10 are presented.

In the case of lightning discharge, the geometric capacitance C_0 may be regarded as a plane capacitance between the cloud and the ground. Thunder clouds take on an anvil dome mushroom-like shape. The cloud–ground system may hence be simplified to a cylindrical plane capacitance, considering a circular base of diameter H for the cloud

$$C_0 = \frac{\varepsilon_0 \pi H^2}{4D}. \tag{7.71}$$

The leader channel is assumed to be in LTE. Its physical background has been established by the Les Renardières group [18].

Rather than computing the length of the first corona from equation (7.26), it has to be set to an initial value varying between 50 and 100 m. The distributed-circuit parameters of the discharge channel are computed using the set of equations presented in section 7.6. However, it should be emphasized that lightning channel recordings reveal diameter values of a few centimetres [31]; consequently, an initial value of $r_0 = 0.5$ cm was found to be appropriate for the simulations [32, 33]. For lightning discharge simulation, a value of $D_\mathrm{f} = 100$ km was acceptable [26, 27, 32, 33].

Another important change to be considered is the mean field in the uncrossed gap E_Z. The simplification for long air gaps (equation 7.41) may no longer be valid. Instead, the following relationship was preferred [32, 33].

$$E_Z = \frac{U_\mathrm{c}(t) - U_\mathrm{i}}{D}, \tag{7.72}$$

where ΔU represents the voltage dropped along the leader channel.

The electric field strength for air breakdown E_0 that is used in the application of the leader inception criterion has to be corrected for variations in altitude z (in km) and air pressure c (in kPa).

The application of such a correction is particularly important when considering lightning protection systems installed at high altitudes [34]. This variation is modelled using a simplified linear approximation [34, 35]:

$$E(Z) = E_0(1 - 0.09Z). \tag{7.73}$$

Changes of pressure with altitude are well approximated using

$$c(Z) = c_0 \exp(-Z/8), \tag{7.74}$$

where c_0 is the standard atmospheric pressure of 1.013 kPa and Z is the altitude (km) [34, 35]. Humidity corrections change E_0 by up to 5% and hence can be ignored [34].

In any case, humidity variations are governed almost entirely by the prevailing weather conditions during the thunderstorm.

References

[1] Les Renardières group 1977 *Electra* **53** 31

[2] Les Renardières group 1986 *Proc. IEEE A* **133** 395

[3] Garcia H and Hutzler B 1974 Electrical breakdown in long air gaps. The final jump *IEEE Third Conf. on Gas Discharges* (London, UK, 9–12 September) p 206

[4] Gallimberti I 1979 *J. Physique* **40** 193

[5] Klewe R C, Waters R T and Jones B 1974 Breakdown phenomena in long air gaps under switching impulse conditions *IEEE Summer Meeting* paper 74-CH 0910 PWR

[6] Le Roy G, Gary C, Hutzler B, Lalot J and Dubanton C 1984 *Les Propriétés diélectriques de l'Air et les Très Hautes Tensions (Collection de la Direction des Études et Recherches d'Électricité de France* vol 51) (Paris: Eyrolles)

[7] Hutzler B and Hutzler D 1982 Modélisation de l'amorçage des grands intervalles d'air *Bull. Dir. Etud. Rech.* **4** 11

[8] Bondiou A and Gallimberti I 1994 *J. Phys. D: Appl. Phys.* **27** 1252

[9] Les Renardières Group 1974 *Electra* **35** 49

[10] Rizk F A M and Vidal F 2008 *IEEE Trans. Power Deliv.* **23** 296

[11] Rizk F A M 1989 *IEEE Trans. Power Deliv.* **4** 596

[12] Vidal F, Gallimberti I, Rizk F A M, Johnson T W, Bondiou-Clergerie A, Comtois D, Kieffer J-C, La Fontaine B, Mercure H P and Pépin H 2003 *IEEE Trans. Plasma Sci.* **30** 1339

[13] Kosztaluk R 1985 *High Voltage Test Techniques* vol 1 (Warsaw: Wydawnictwa Naukowo-Techniczny) in Polish

[14] Aguet M and Ianoz M 1987 *Haute Tension (Traité d'Électricité)* (Paris: Dunod)

[15] Ortega P, Domens P, Dupuy J, Gibert A, Ruhling F and Heilbronner F 1991 Long air gap discharges under non-standard positive impulse voltages. Part 2: physical interpretation *7th Int. Symp. on High Voltage Engineering* (*Dresden, Germany, 26–30 August*) p 105

[16] Giao N T 1995 *IEEE Electr. Insul. Mag.* **2** 5

[17] Fofana I and Beroual A 1996 *J. Phys. D: Appl. Phys.* **29** 691

[18] Beroual A 1993 *J. Appl. Phys.* **73** 4528

[19] Beroual A 2003 universal dynamic model of discharge propagating in air, liquids, solid/liquid interfaces and polluted *13th Int. Symp. on High Voltage Engineering* (*Delft, The Netherlands, August 25–29*)

[20] Fofana I and Béroual A 1997 *J. Phys. D: Appl. Phys.* **30** 1653–67

[21] Ortéga P 1992 Comportement diélectrique des grands intervalles d'air soumis 'a des ondes de tension de polarité positive ou négative *PhD Thesis* Université de Pau et des Pays de l'Adour

[22] Hutzler B 1988 *Bull. Dir. Etud. Rech.* **3** 31

[23] Hutzler B 1988 Comparaison foudre—étincelle de laboratoire *Texte de Communications, SE, Groupe: Décharges Électriques* p 21

[24] Fofana I and Beroual A 1995 Modelling of the leader and the return stroke currents with an equivalent electrical network *9th Int. Symp. High Voltage Engineering* (*Graz Austria August, 28–September 1*) paper 2141

[25] Fofana I and Beroual A 1995 *J. Phys. D: Appl. Phys.* **28** 303

[26] Fofana I and Beroual A 1996 *IEEE Trans. Dielectr. Electr. Insul.* **3** 273

[27] Ratnamahilan P 1993 *IEEE Trans. Magn.* **29** 1839

[28] Taudière I 1988 Analyse du rayonnement électromagnétique VHF-UHF de l'éclair. Application à un modèle d'avancée du précurseur négatif *PhD Thesis* Université de Paris VI

[29] Domens P 1987 Contribution à l'étude des décharges électriques dans les grands intervalles d'air *PhD Thesis* Université de Pau

[30] Gary C 1995 *La Foudre: des Mythologies Antiques à la Recherche Moderne* (Paris: Edition Masson)

[31] Uman M A 1987 *The Lightning Discharge (International Geophysics Series) vol 39* (New York: Academic)

[32] Fofana I and Beroual A 1996 A positive lightning stepped leader model based on the similarity with long laboratory sparks *23rd Int. Conf. Lightning Protection* (*Firenze, Italy, September 23–27*) p 393

[33] Fofana I, Ben Rhouma A, Beroual A and Auriol P 1998 *IEEE Proc. Gen. Trans. Distrib.* **145** 395

[34] D'Alessandro F and Gumley J R 2001 *J. Electrostatics* **50** 279

[35] Fofana I and Beroual A 2004 A quantitative study of lightning striking distance factors *Proc. 27th Int. Conf. Lightning Protection (Avignon, France, 13-16 September)* p 421

IOP Publishing

Discharge in Long Air Gaps
Modelling and applications
A Beroual and I Fofana

Chapter 8

Modelling the negative discharge in long air gaps

8.1 Introduction

While the positive discharge in long air gaps has been the subject of numerous experimental [1–11] and theoretical studies [12–20], only a few quantitative results have appeared concerning the negative discharge; the literature related to negative discharge remains very poor [10, 21–23] compared to its positive counterpart. Also, many theoretical models that describe positive discharge are available, yet proposed models to describe negative discharge are practically non-existent due to the complexity of the mechanisms involved.

This chapter presents a model that enables the dynamic simulation of the successive phases of negative discharge (the initiation of the first corona, the pilot system, the negative and space leaders, the junction of leaders, etc), including criteria for instabilities and arrest conditions. This model uses a distributed *RLC* circuit, the parameters of which vary with time according to the discharge characteristics and geometry (*R*, *L*, and *C* being the resistance, the inductance, and the capacitance, respectively). It is based on previous models related to positive discharge in long air gaps and lightning [14–18] and it takes into account important parameters, such as electrode geometry, impulse voltage shape, and some atmospheric conditions.

8.2 The development of a negative discharge

As indicated in chapter 4, the negative discharge is a non-continuous phenomenon which progresses through regular and discrete 'steps'. A typical propagation sequence of such a discharge can be schematically described as follows (figure 8.1) [21]:

(1) The inception of the first corona-streamers at the HV electrode.
(2) A few microseconds later, the extinction of the first corona-streamers is followed by a short dark period before the inception and development of a bipolar system called the pilot. This latter is made of two coronas of

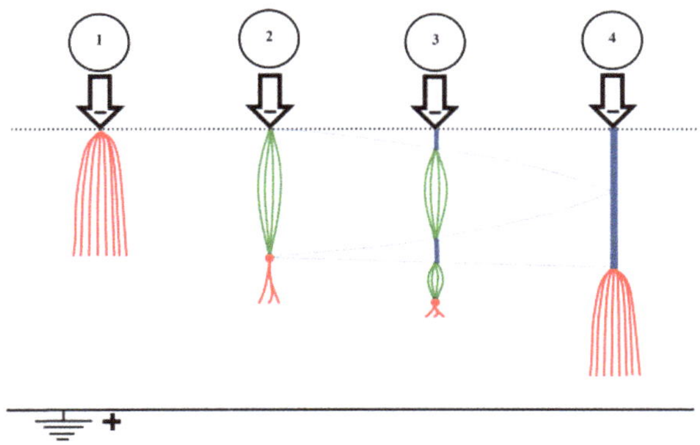

Figure 8.1. Simplified schema of negative discharge propagation. Reproduced with permission from [24].

opposite polarity: the positive corona develops toward the HV electrode while the negative corona propagates toward the grounded plane electrode. The pilot system repeats several times with a 1–2 μs period by progressing in the gap. At regular intervals (1–2 μs), new similar pilots develop from the tip of the preceding ones that progress in the gap.

(3) The development of a space leader from a 'space stem', actually a previous pilot inception point. The space leader progression, as a bidirectional discharge toward the anode (positive end) and the cathode (negative end), is followed by the inception of a negative leader from the negative electrode.

(4) Finally, the junction between both leaders (space leader and negative electrode leader) produces an abrupt elongation of the channel. This junction is accompanied by a strong re-illumination of the discharge channel while a new corona takes place at the head of the discharge channel. Thus the stepwise propagation mechanism is completed. The entire phenomenon resumes and will be repeated until the grounded plane is reached. A schematic representation of negative propagation can be found in figure 8.2.

8.3 Theoretical background

8.3.1 Corona initiation and development

The initiation and development mechanisms of the first corona under a negative voltage are similar to those of positive polarity (see chapters 3 and 7). The electric field required for the inception of the corona may be calculated using Peek's formula (7.24), while the electric field E_p at the HV electrode is computed using the hyperboloidal approximation described in section 7.4.

Taudière has shown that the stability field is equivalent to the average field along the streamer extension [25]; therefore, if the average field is fixed, the length of the negative corona l_s is directly given by the graphical construction (figure 8.3), i.e. the geometrical method, by considering the electric field within the corona region as constant [25]. Hence, l_s is determined using the electrostatic potential distribution before and after corona development.

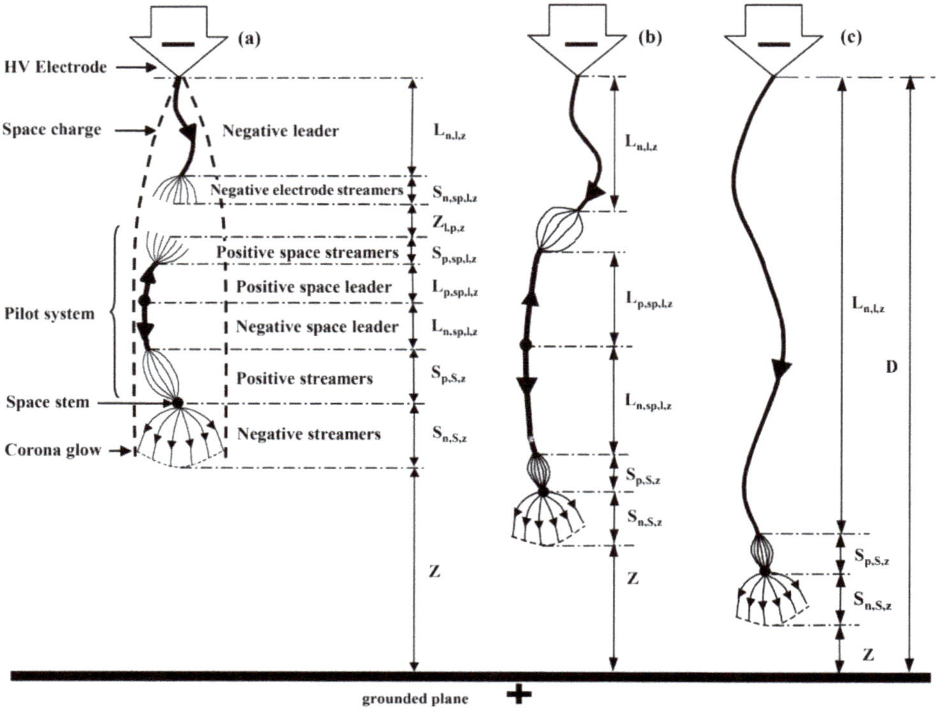

Figure 8.2. A schematic representation of negative discharge propagation. $L_{n,l,z}$: negative electrode leader; $S_{n,sp,l,z}$: negative space streamers; $S_{p,sp,l,z}$: positive space streamers; $L_{p,sp,l,z}$: positive space leader; $L_{n,sp,l,z}$: negative space leader; $S_{p,S,z}$: positive streamers; $Z_{n\text{-}p}$: streamer zone; $S_{n,S,z}$: negative streamers; Z: uncrossed distance; and D: electrode gap. Discharge channel before (a), during (b), and after (c) the junction between the positive space leader and negative electrode leader: re-illumination. Reproduced with permission from [24].

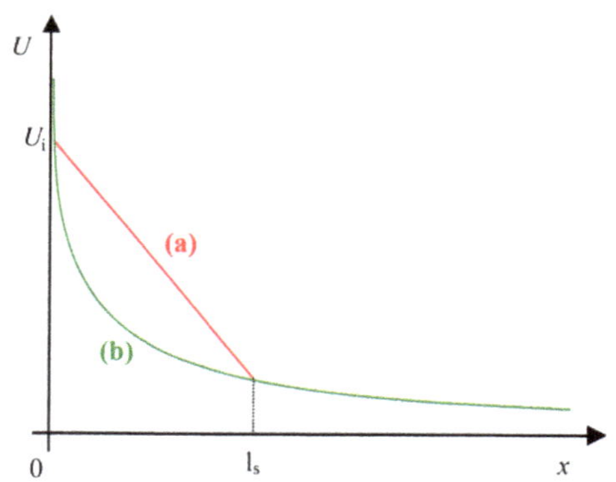

Figure 8.3. Voltage distribution before (a) and after (b) the development of the negative corona-streamers. Reproduced with permission from [24].

8.3.2 Development of the pilot system

Both positive and negative coronas are initiated once the criterion related to the electric field at the head of the streamers is satisfied. The electric field at the head of the streamers is calculated in the same manner as for the initiation of the first corona:

$$E_{s,\text{head}} = \frac{2V_{s,j}}{r_s \ln\left(\dfrac{4Z}{r_s}\right)}. \qquad (8.1)$$

Z is the gap uncrossed by the discharge; $V_{s,j}$ is the voltage at the streamer's head; r_s is the negative streamer's radius estimated analytically from the relationship giving the electric field of stability [13]. After the extinction of these newly formed coronas, another relaxation process takes place along the pilot system and so on.

8.3.3 Criteria for electrode leader growth

The growth of the 'negative electrode leader' is mainly governed by the field $E_{l,p}$ at its tip. This field can be calculated by assuming the negative leader tip to be an equivalent electrode (section 7.4). Depending on the value of $E_{l,p}$ at this electrode surface, the leader channel may or may not propagate. If $E_{l,p}$ is higher than the critical field E_c, the leader channel will propagate through the gap to keep the corona effect alive. Otherwise, the leader channel progressively cools and finally disappears, unless the applied voltage characteristics are able to re-activate it. $E_{l,p}$ can be estimated using a hyperboloidal approximation [12]:

$$E_{l,p} = \frac{2V_j}{r_e \ln 4\left(\dfrac{D - L_{n,l,z}}{r_e}\right)}. \qquad (8.2)$$

V_j is the voltage at the electrode leader tip, r_e is the radius of the equivalent hyperboloid electrode; this latter can be computed using the empirical relationship proposed in the literature [12]

$$r_e = \frac{V_j}{V_{j0}} \frac{0.25}{1 + \dfrac{3.5}{L_{n,l}}} + r_{10} \exp\left(\frac{L_{n,l,z}}{-13.6 r_{10}}\right), \qquad (8.3)$$

where r_{10}, $L_{n,l}$, and $L_{n,l,z}$ are the initial equivalent radius after the development of the first corona, and the leader's effective and axial lengths, respectively. V_{j0} is the value of the voltage V_j that one would have to apply to the unbridged gap to make the field on the equivalent electrode surface equal to the critical field E_c [12]. Thus, from equations (8.2) and (8.3), we have:

$$V_{j0} = \frac{E_c r_e}{2} \ln 4\left(\frac{D - L_{n,l,z}}{r_e}\right) \qquad (8.4)$$

and

$$r_e = \frac{0.25}{1 + \frac{3.5}{L_{n,l}}} + r_{10} \exp\left(\frac{L_{n,l,z}}{-13.6 r_{10}}\right). \tag{8.5}$$

Note that the electric field can be calculated thanks to numerical techniques. However, in the rest of the chapter, the above analytical expression will be used.

8.3.4 Development of electrode and space leaders

The propagation of electrode and space leaders is related to the power P_j (or energy W_j) injected at the head of each leader by the corona and pilot systems (see section 7.4). This power is computed from the equivalent electric circuit RLC that is described in section 8.4. On the other hand, since the negative discharge propagates by steps, the leader propagation is represented by a succession of segments of length $dL_{n,l}$ (where $dL_{n,l} = v dt$, v and dt being, respectively, the leader propagation velocity and the computation time step) making a random angle φ with the gap axis, that is $dL_{n,l,z} = dL_{n,l} \cos\varphi$, $dL_{n,l,z}$ being the leader step's axial length.

The negative electrode leader propagation velocity can, therefore, be calculated using the relationship proposed for the positive discharge [17, 18, 20, 26]

$$v_{j,n,l}(t) = \left(\frac{2\beta_{n,l}}{\rho\pi r_l^2}\frac{dW_j}{dt}\right)^{1/3} \tag{8.6}$$

or

$$v_{j,n,l}(t) = \left(\frac{2\beta_{n,l}}{\rho\pi r_l^2} dP_j\right)^{1/3}, \tag{8.7}$$

where ρ is air density (ρ is set to 0.15 kg m^{-3}) and r_l is the leader tip radius at the time, t_0, when it is formed. $\beta_{n,l}$ is the fraction of the available energy used for the negative electrode leader displacement ($0 < \beta_{n,l} \leqslant 1$) . Its value is estimated to be about 0.1.

In the same way, the velocities of both positive and negative spatial leaders ($v_{j,p,sp,l}$ and $v_{j,n,sp,l}$, respectively) were computed by replacing $\beta_{n,l}$ with $\beta_{p,sp,l}$ and $\beta_{n,sp,l}$, respectively. The values of $\beta_{p,sp,l}$ and $\beta_{n,sp,l}$ are much smaller than $\beta_{n,l}$. In our modelling, the lengths of the spatial leaders are supposed to represent a small percentage of the negative electrode leader.

The erratic propagation of the discharge, defined by the random angle φ with the gap axis considered in the modelling, is the same as for the positive discharge (see section 7.4).

8.3.5 Voltage distribution

Before the junction, the potential drop across the gap can be defined as

$$U_c(t) = (E_{n,l}L_{n,l} + E_{n,s}L_{n,s}) + \left(E_{p,sp,l}L_{p,sp,l} + E_{p,sp,S}S_{p,sp,l}\right) + \left(E_{p,sp,l}L_{n,sp,l} + E_{n,sp,S}S_{n,sp,l}\right) + \left(E_{Z,l-p}Z_{l-p}\right) + (E_Z Z), \quad (8.8)$$

where:

- $E_{n,l}$, $E_{p,sp,l}$, and $E_{n,sp,l}$ are the mean fields within the negative electrode leader, the positive spatial leader, and the negative spatial leader, respectively;
- $E_{n,S}$, $E_{p,sp,S}$, and $E_{n,sp,S}$ are the mean fields within the streamer zone of the negative electrode leader, the streamer zone of the positive spatial leader, and the streamer zone of the negative spatial leader, respectively;
- $L_{n,l}$, $L_{p,sp,l}$, and $L_{n,sp,l}$ are the lengths of the negative electrode leader, the positive spatial leader, and the negative spatial leader, respectively;
- $S_{n,S}$, $S_{p,sp,l}$, and $S_{n,sp,l}$ are the lengths of the streamer zone of the negative electrode leader, the streamers zone of positive spatial leader, and the streamer zone of negative spatial leader, respectively;
- $E_{Z,l\text{-}p}$ and E_Z are the mean fields within the gap between the streamer zones of both the negative electrode leader and the positive spatial leaders, and the uncrossed electrode gap (i.e. between the head of the whole discharge channel and the opposite electrode);
- $Z_{l\text{-}p}$ and Z are the axial lengths between the streamer zones of both the negative electrode leader and the negative spatial leaders, and the uncrossed electrode gap (see figure 8.2). $E_{n,l}$ can be determined using non-LTE considerations [8].

According to experimental observations [10, 21], the streamers at the head of both the negative electrode and the positive spatial leaders are not visible, probably due to their high resistivity (fewer ionized zones). In our modelling, some assumptions were made concerning these streamer zones and the gap between them before the junction. This consists in extrapolating these streamer zones to parts of the associated leaders and replacing their lengths by an equivalent one representing a small percentage of the considered leader channels.

After the junction, the electric field E_Z of the uncrossed gap is computed using equation (7.41).

The length of the leader corona-streamers can be deduced from equations (7.40) and (8.8).

8.3.6 Attachment process

Recall that the random propagation of the leader and dispersion in the inception voltage are responsible for the dispersion of voltages and times to re-illumination. The propagation of the head of the positive leader towards the cathode leads to the junction with the negative electrode leader propagating in the opposite direction. The distribution of the potential at the end of the re-illumination is calculated geometrically. Neglecting the electric field in the channel of the new leader, the

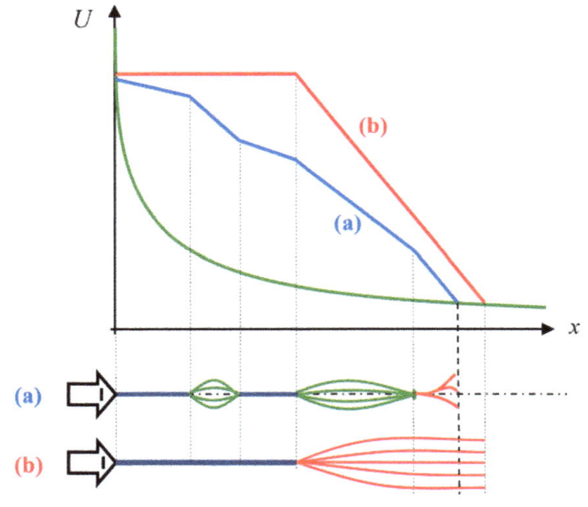

Figure 8.4. Potential distribution before (a) and after (b) the junction between the positive space leader and the electrode leader. Reproduced with permission from [24].

injected current will be high enough to allow thermalization of the channel leading to a significant decrease in the internal field [22]. As a consequence, the resistance per unit length of the leader channel is neglected.

At the tip of the new leader, a negative corona will be initiated. The length of this corona is also geometrically determined [13] (figure 8.4).

8.3.7 Discharge trajectory and dispersion

To take into account the hazardous propagation of the discharge, we used the same probabilistic law established for the long positive discharge by Hutzler and Hutzler [12] (see section 7.4).

8.3.8 The final jump

Knowing that three different types of final jump may occur (see chapter 4), only the one starting when the negative streamers reach the plane (type A) is considered [21]. In this case, the arrival of the negative streamers at the plane may occur either suddenly after an elongation of the channel, or during the propagation of the leader and of the associated space stem. Thus, the final jump appears to be strongly linked, through its initial conditions, to the propagation mode of the discharge and, consequently, to the experimental conditions. The height of the final jump is determined by the distance between the negative leader tip and the plane, at the time when the negative streamers arrive at the plane. In almost all cases, a positive leader will grow from the grounded plane electrode. In most cases, the length of the positive leader is lower than or equal to the half-height of the final jump. Finally, the development of the final jump is directly linked to the mode of leader propagation. Its spatial and temporal characteristics are, therefore, dependent on the gap length and voltage characteristics.

The transition from the free development to the final jump phase is initiated when the value of the total axial length of the negative discharge $L_{t,z}$ (after the junction phase) exceeds the height of the final jump L_{fj}, which depends only on the voltage at the leader tip when the first streamers reach the plane [2]. The value of L_{fj} is considered to be similar to that established for positive long discharges (see section 7.4)

The initiation of this phase leads inevitably to breakdown unless the applied voltage drops substantially. The conditions for stop and go at this stage are similar to those described in section 7.4, concerning positive polarity.

8.4 Distributed-circuit-based modelling

The distributed-circuit-based model is based on the same concept used for positive polarity. This consists of representing the discharge channel by the equivalent electrical circuit depicted in figure 8.5, where R_0 is the plane electrode resistance and n the number of leader steps. A value of $R_0 = 100\ \Omega$ is assumed [14–18], while C_0 is taken as a half-spherical capacitance, estimated from a spherical approximation (the gap geometry is assimilated to two concentric spheres of radius R_p and $(R_p + D)$). C_0 is computed from equation (7.63).

Starting from the first corona, a new *RLC* cell associated with the new pilot is created. The discharge development is illustrated by the diagrams given in table 8.1.

At each time, one can write the electrical equations governing the leader propagation according to the equivalent electrical circuit:

$$L_j \frac{\mathrm{d}ij}{\mathrm{d}t} = V_{j-1}(t) - V_j(t) - R_j i_j(t) \tag{8.9}$$

$$C_j(z) \frac{\mathrm{d}V_j(t)}{\mathrm{d}t} = i_j(t) \tag{8.10}$$

and

$$i_{j-1}(t) - i_j(t) = C_{j-1} \frac{\mathrm{d}V_{j-1}(t)}{\mathrm{d}t}. \tag{8.11}$$

The discharge current, the charge injected in the gap, and its energy and power are computed using equations (7.54), (7.55), (7.35), and (7.36), respectively.

8.4.1 The resistance

The leader corona

The resistance per unit length of leader corona is calculated assuming that the leader corona behaves as a leader portion, i.e.

$$\lambda_s = \frac{E_s}{I_s}. \tag{8.12}$$

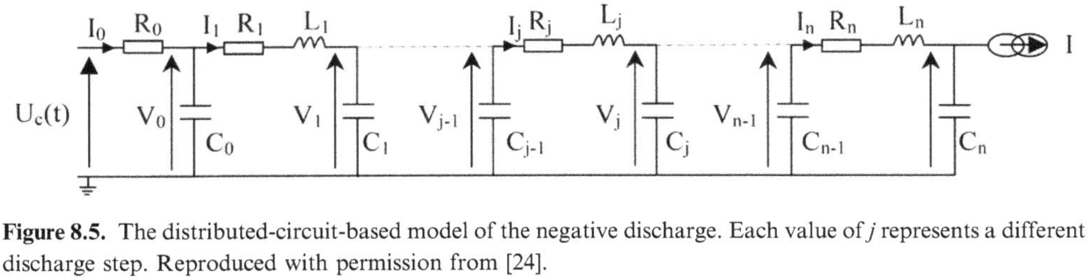

Figure 8.5. The distributed-circuit-based model of the negative discharge. Each value of *j* represents a different discharge step. Reproduced with permission from [24].

Table 8.1. Simulation of some corona-like discharge processes.

<table>
<tr><th>Actual discharge</th><th>Equivalent electric circuit</th></tr>
<tr><td>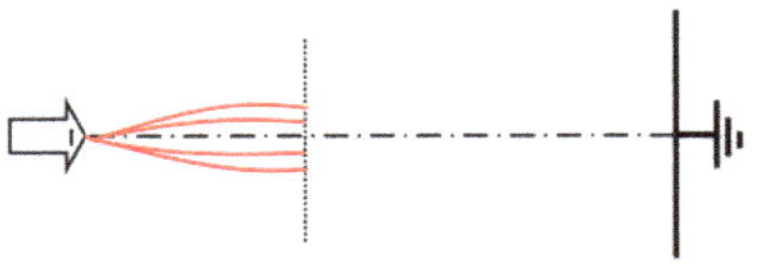
Negative corona-streamers</td><td>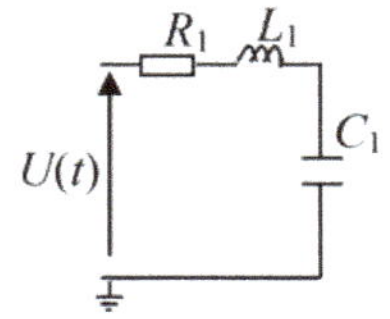
</td></tr>
<tr><td>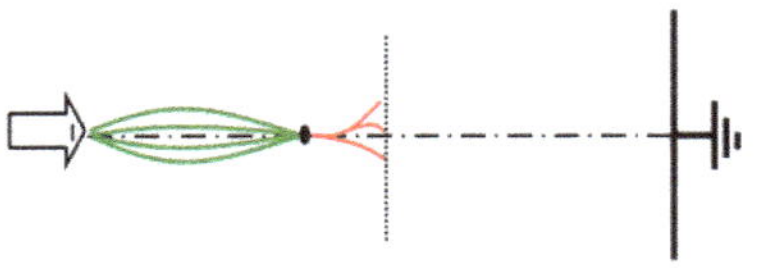
Pilot : constituted by positive and negative corona-streamers</td><td>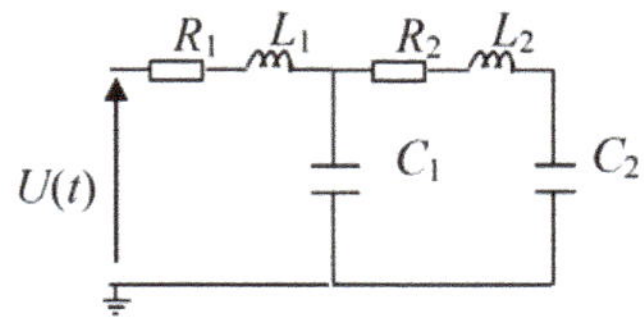
</td></tr>
<tr><td>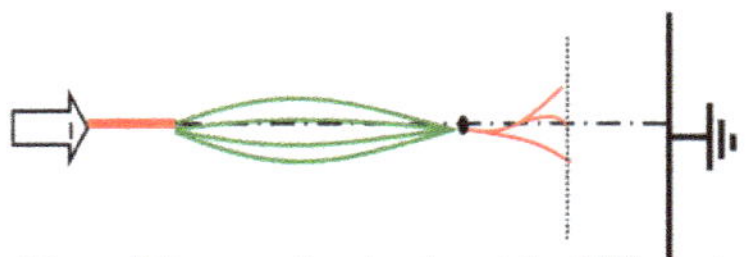
Apparition of the negative leader at the HV electrode.</td><td>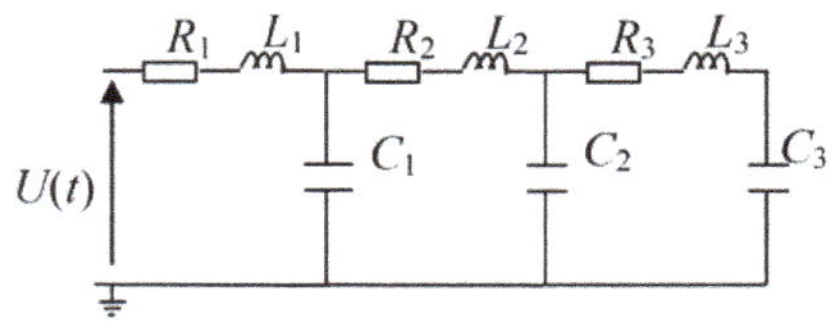
</td></tr>
</table>

E_s is the electric field within the streamer zone taken at 15 kV cm^{-1} [12]. I_s is the sum of the current flowing through the streamers.

The resistance and inductance and capacitance

Since LTE assumptions do not apply during leader propagation, the channel of the negative leader can, therefore, be considered as a resistive electrode, the properties of which are controlled by Ohmic heating and gas dynamic expansion. The resistance per unit length of both leaders is computed using the same approach described for the positive leader. Figure 7.8 and the set of equations (7.38), (7.56)–(7.58), are to be considered in the computation of the resistance per unit length of the leaders. The resistance per unit length of the leaders during the final jump is computed from equation (7.59).

As far as the return stroke phase is concerned, it is known [4] that plasma can be considered to be in LTE within the spark: the gas molecules are fully dissociated, and the temperature is uniform throughout the channel. The simulation of this phase is performed using the same concept described in section 7.6.

While the inductance of the leaders is computed using the simplification highlighted in section 7.6, the corona leader capacitance together with the capacitance of the uncrossed gap can be computed using the set of equations (7.64)–(7.70) described in section 7.6.

8.5 General description of computation steps

The impulse voltage wave shape is simulated using the same approach as described in section 7.3. The only difference is the polarity of the input voltage U_0.

The flowchart given in figure 8.6 provides an overview of the computer program developed to simulate the negative discharge. T_M represents the maximum duration specified for the total discharge development in a given configuration (electrode gap and geometry, applied voltage, and atmospheric conditions). Since the model is a self-consistent one, the only input data are the geometry of the electrodes, the voltage waveform, and the ambient air pressure and temperature.

The discharge time base is divided into steps d*t*, (lower than the arc time constant) starting from $t = 0$. Note that d*t* is appropriately chosen to have the best fitting of experimental data, i.e. very small compared to the scale of the *RC* cell's time constant. Depending on the air gap length, d*t* has been varied between 10^{-8} to 10^{-6} s.

At each time step, the applied voltage $U_c(t)$ is calculated and the inception conditions (especially the critical electric fields) of any step of the discharge are tested. The simulation starts with the first corona inception test. If the criterion is satisfied, the first corona length S_0 is compared to L_{fj} to check if the final jump is reached or not. At each time base step, the axial length of the negative leader $L_{n,l,z}$ is compared to L_{fj}. Other tests are performed in the different loops depicted in the flow chart (figure 8.6). Among the different tests that one has to perform that of the stem velocity. If this latter is negative and the time exceeds a reference time T_M, depending on the applied wave shape, then there is withstand. Otherwise, the electric field at the negative leader head E_p and the critical field E_c corresponding to the unbridged gap are computed. If $E_p > E_c$, the leader propagation starts. At this time, if there were no spatial leaders, then these later will be initiated. The different lengths of discharges are calculated from the instantaneous velocity and time base step. If the spatial leaders were already initiated, then a test is performed to check if the negative electrode leader joins the upward positive spatial leader leading to the re-illumination. The different electrical parameters are then updated and the simulation is re-looped up to the final jump or withstand.

The internal conditions of the discharge channel (radius, gas number density, electric field, electrical parameters, and so on) and the unbridged gap capacitance are calculated. Then we compute the temporal evolution of the currents, the potential gradient within the leader channel, the power injected into the gap, and the channel's thermal radius.

Simulations of an image converter working in streak or frame mode, the discharge propagation velocities, and a trajectory of the discharge plotted in real time may also be obtained. The results are compared with experimental data.

8.6 Extension to a very long air gap: negative lightning

A mathematical model of negative lightning discharge was developed based on the idea that a substantial similarity exists between the lightning phenomenon and the discharge in large air gaps. The same concept was therefore used to simulate lightning discharge. In section 7.8, important changes that allowed the extension of the long spark model to lightning are described.

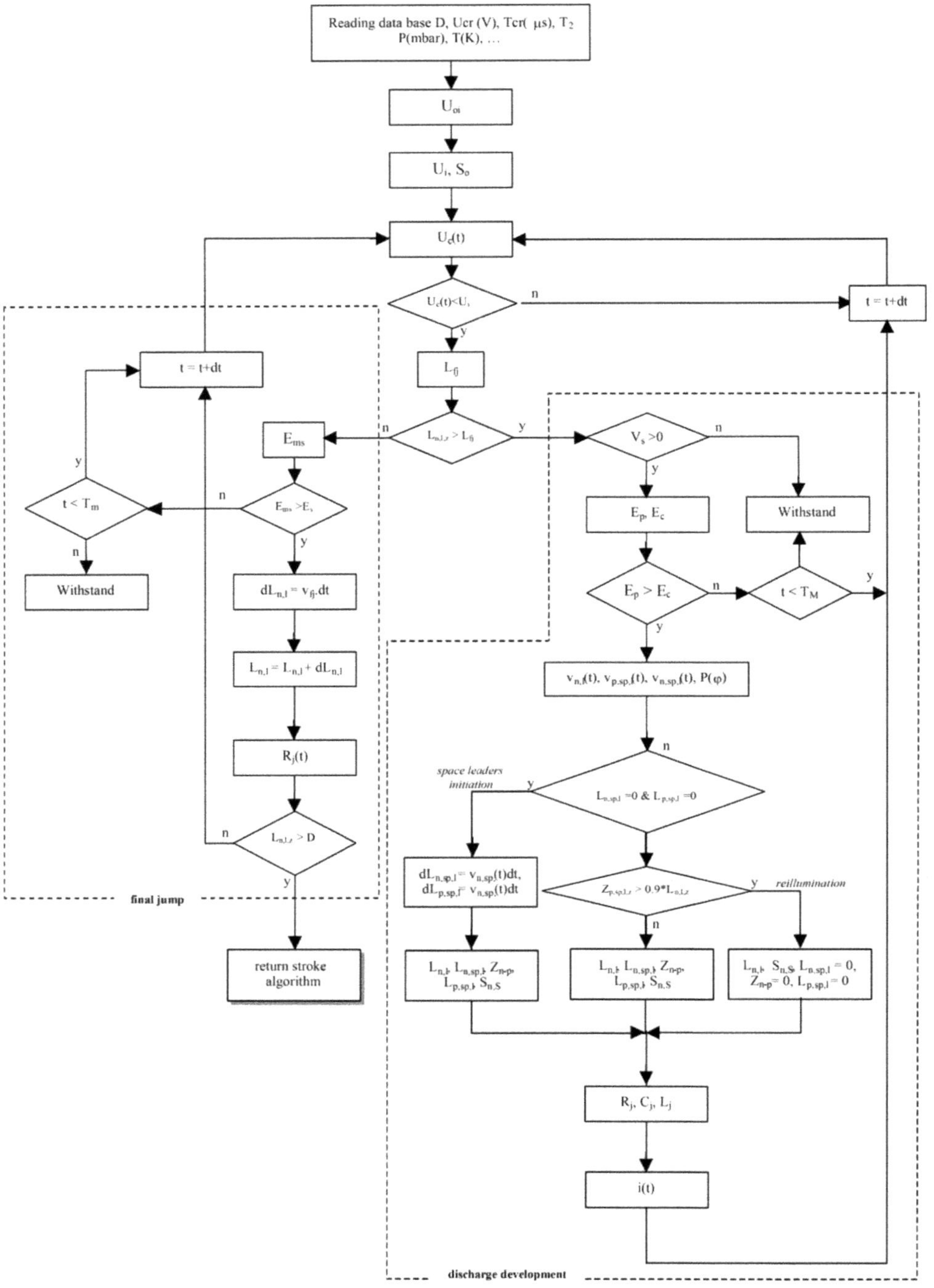

Figure 8.6. Block diagram showing a negative discharge process. Reproduced with permission from [24].

References

[1] Schmitt A and Deflandre T 1999 Les surtensions et les transitoires rapides de tension en milieu industriels et tertiaire *Collection Direction des Etudes et Recherches d'Electricité de France* (Paris: EDF)

[2] Garcia H and Hutzler B 1974 Electrical breakdown in long air gaps. The final jump *IEEE Third Conf. on Gas Discharges* (*London, UK, 9–12 September*) p 206

[3] Klewe R C, Waters R T and Jones B 1974 Breakdown phenomena in long air gaps under switching impulse conditions *IEEE Summer Meeting* paper 74-CH 0910 PWR

[4] Gallimberti I 1979 *J. Physique* **40** 193–205

[5] Le Roy G, Gary C, Hutzler B, Lalot J and Dubanton C 1984 *Les Propriétés diélectriques de l'Air et les Très Hautes Tensions* (*Collection de la Direction des Études et Recherches d'Électricité de France* vol 51) (Paris: Eyrolles)

[6] Bazelyan E M, Raizer Yu P and Aleksandrov N L 2007 *J. Phys. D: Appl. Phys.* **40** 4133

[7] Les Renardières Group 1977 *Electra* **53** 31–153

[8] Les Renardières Group 1986 *Proc. IEEE A* **133** 395

[9] Domens P 1987 Contribution à l'étude des décharges électriques dans les grands intervalles d'air *PhD Thesis* Université de Pau et des Pays de l'Adour, France

[10] Ortéga P 1992 Comportement diélectrique des grands intervalles d'air soumis à des ondes de tension de polarité positive ou négative *PhD Thesis* Université de Pau et des Pays de l'Adour, France

[11] Giao N T 1995 *IEEE Electr. Insul. Mag.* **2** 5

[12] Hutzler B and Hutzler D 1982 *Bull. Dir. Etud. Rech* **4** 11

[13] Bondiou A and Gallimberti I 1994 *J. Phys. D: Appl. Phys.* **27** 1252

[14] Fofana I and Beroual A 1995 Modelling of the leader and the return stroke currents with an equivalent electrical network *9th Int. Symp. High Voltage Engineering* (*Graz, Austria, August 28–September 1*) paper 2141

[15] Fofana I and Beroual A 1995 *J. Phys. D: Appl. Phys.* **28** 303

[16] Fofana I and Beroual A 1996 *IEEE Trans. Dielectr. Electr. Insul.* **3** 273

[17] Fofana I and Beroual A 1996 *J. Phys. D: Appl. Phys.* **29** 691

[18] Fofana I and Beroual A 1997 *J. Phys. D: Appl. Phys.* **30** 1653

[19] Goelian N, Lalande P, Bondiou-Clergerie A, Bacchiega G L, Gazzani A and Gallimberti I 1997 *J. Phys. D: Appl. Phys.* **30** 2441

[20] Beroual A 2003 universal dynamic model of discharge propagating in air, liquids, solid/liquid interfaces and polluted *13th Int. Symp. High Voltage Engineering* (*Delft, The Netherlands, August 25–29*)

[21] Les Renardières Group 1981 *Electra* **74** 67

[22] Bacchiega G L, Gazzani A, Bernardi M, Gallimberti I and Bondiou-Clergerie A 1994 Theorical modelling of the laboratory negative stepped leader *16th Int. Aerospace and Ground Conf. on Lightning and Static Electricity* (*Mannheim, Germany, May 24–25*)

[23] Beroual A 2004 *REE Rev. Electr. Electron.* **8** 36

[24] Rakotonandrasana J H *et al* 2008 *J. Phys. D: Appl. Phys.* **41** 105210

[25] Taudière I 1988 Analyse du rayonnement électromagnétique VHF–UHF de l'éclair—Application à un modèle d'avancée du précurseur négatif *PhD Thesis* Université de Paris VI

[26] Beroual A 1993 *J. Appl. Phys.* **73** 4528

IOP Publishing

Discharge in Long Air Gaps
Modelling and applications
A Beroual and I Fofana

Chapter 9

Applications of the model developed for positive discharge in long air gaps

9.1 Introduction

In this chapter, our principal aim is to show that the use of the self-consistent dynamic mathematical model we developed (see chapter 7) enables us to predict some major parameters relevant to long air gap discharge under positive polarity. The case of long air gaps when inserting a floating barrier is also considered. The results predicted by our model are validated against experimental ones reported by different researchers, considering different gap lengths, wave shapes, electrode geometries (with and without a floating barrier), and some atmospheric conditions. We show that the outputs from the model closely follow the experimental results and can predict the critical breakdown voltage or flashover voltage with acceptable accuracy; therefore, the required full-scale testing can be reduced.

9.2 Prediction of the characteristics of long air gap discharges: simulations of some laboratory experiments

The purpose of the simulation of a laboratory experiment lies in the validation it can bring to the modelling. Thus, to make a precise judgement, it is critical to choose a laboratory configuration for which experimental data can be obtained.

The results of simulations for two cases, (a) breakdown and (b) withstanding, are given in figure 9.1, where we can see the shape of the applied voltage, and the leader current and its trajectory via photography (the dashed line represents the final jump phase) of a 5 m point–plane gap subjected to a 200/10 000 μs voltage wave.

For the breakdown case, the final jump phase is marked by the dashed line in the leader trajectory. Notably, the gap voltage drops entirely during the arc phase, as observed experimentally [1].

In order to analyse the possibilities of the model for other wave shapes (an oscillating impulse, for example), a simulation has been performed under the same

doi:10.1088/978-0-7503-1236-3ch9

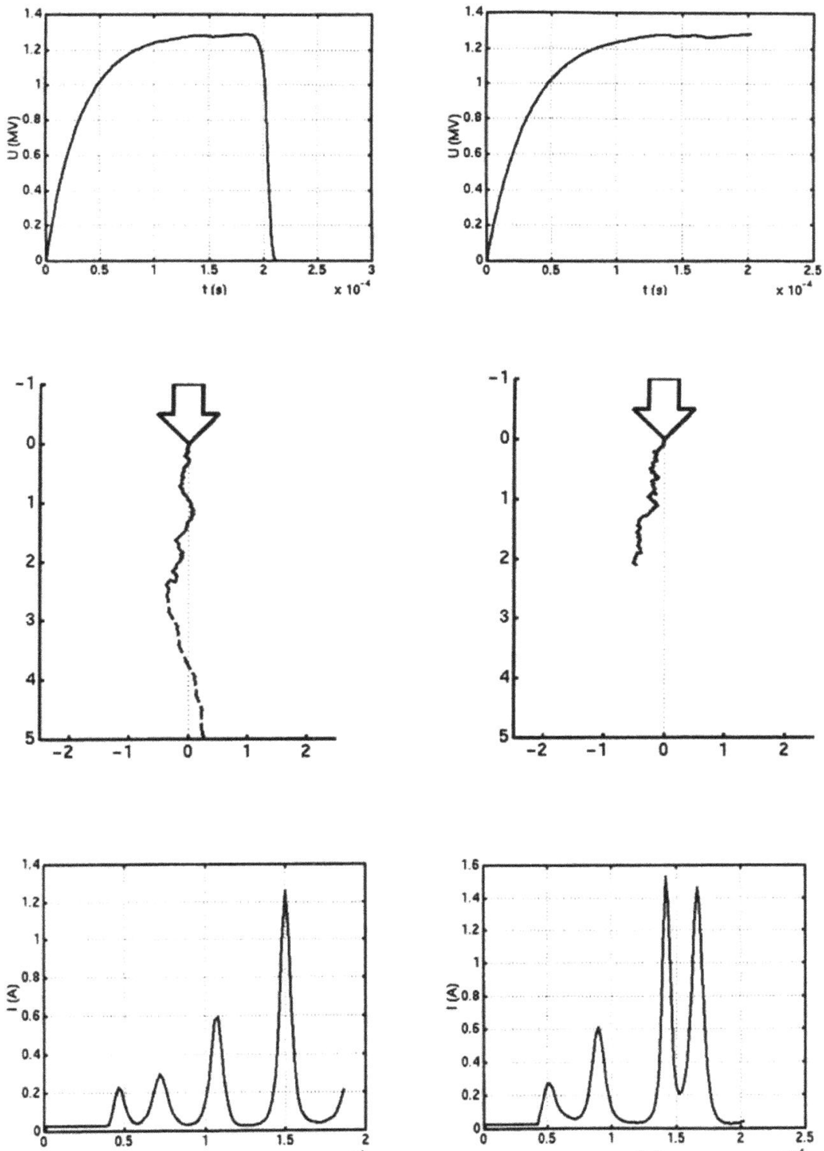

Figure 9.1. A simulation of a point–plane long air gap discharge, showing the leader's trajectory in space and its current. $D = 5$ m, 200/10 000 μs wave, $U_{cr} = U_{50} = 1298$ kV, $P = 997$ mb, $T = 19$ °C and $R_p = 6$ mm. (a) Breakdown and (b) withstanding. Reproduced from [12] with permission.

conditions, as shown in figure 9.2. The frequency of the applied voltage is about 50 kHz. Figures 9.2(a) and 9.2(b) show the shape of the applied voltage and the leader current, respectively. Figures 9.2(c) and 9.2(d) are simulations of image converter working in the streak and frame modes, respectively. Figure 9.2(c) depicts the arrival of the streamers at the plane at the beginning of the final jump phase. In figure 9.2 (d)

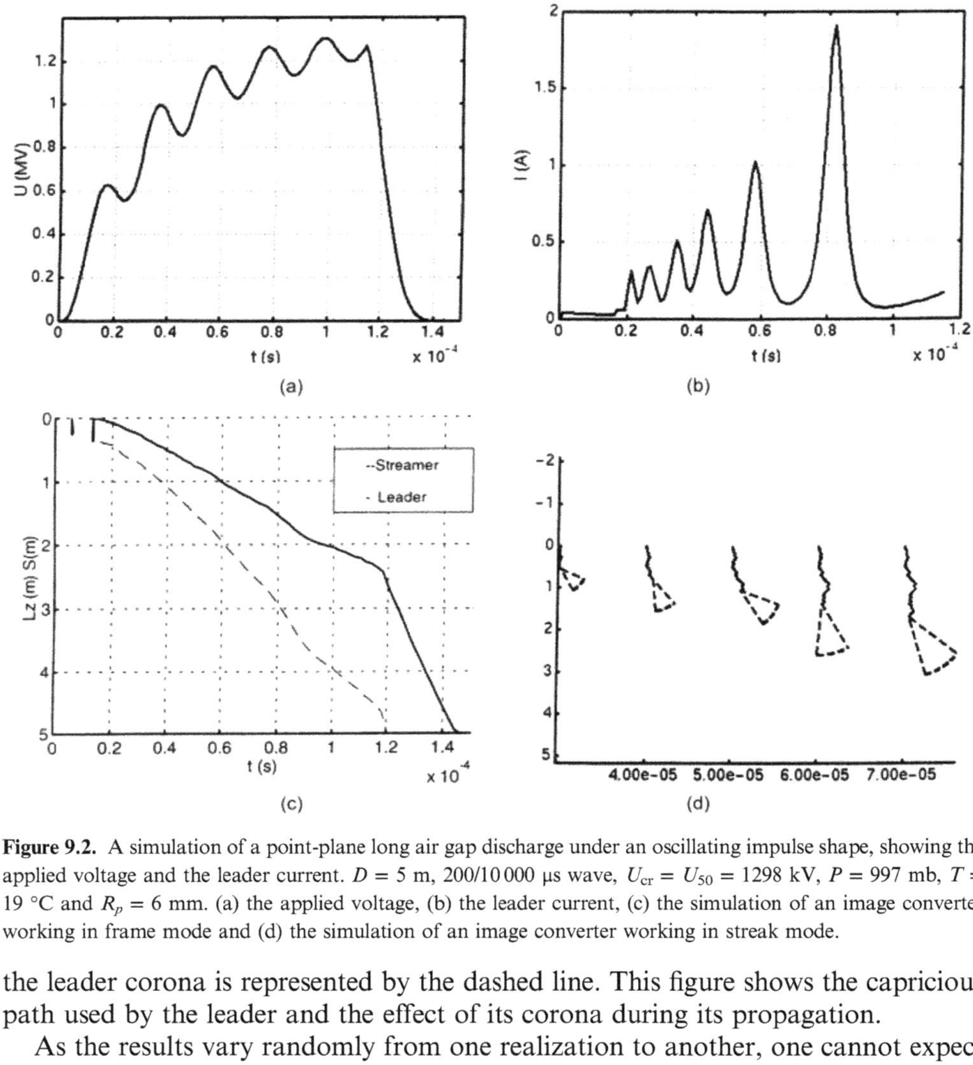

Figure 9.2. A simulation of a point-plane long air gap discharge under an oscillating impulse shape, showing the applied voltage and the leader current. $D = 5$ m, 200/10 000 μs wave, $U_{cr} = U_{50} = 1298$ kV, $P = 997$ mb, $T = 19$ °C and $R_p = 6$ mm. (a) the applied voltage, (b) the leader current, (c) the simulation of an image converter working in frame mode and (d) the simulation of an image converter working in streak mode.

the leader corona is represented by the dashed line. This figure shows the capricious path used by the leader and the effect of its corona during its propagation.

As the results vary randomly from one realization to another, one cannot expect agreement in lesser details between a single run and a single experimental discharge. However, the general characteristics are clearly reproduced by the simulation; for example, the randomness of the leader path and a current consisting of many pulses.

These examples illustrate individual cases. To facilitate the comparison with experimental results, it is desirable to dispose of global results based on a great number of shots. Consider, for example, electrode gaps for which we know the shape of the applied voltage and the corresponding experimental conditions [1]. The global results are compared to the experimental ones for the same number of shots under the same conditions of breakdown (electrode shape, atmospheric conditions, and the applied voltage shape) as shown in table 9.1. In this table T_i, U_i, and T_b are the inception time of the leader, its inception voltage, and the breakdown time, respectively. We observe that all the parameters characterizing the discharge for a

Table 9.1. A comparison of the model results with experimental ones.

Electrode shape	No of shots	D (m)	Impulse shape T_1/T_2 (μs)	U_{cr} (kV) (U_{5o})	Atmospheric conditions P(mb)	T (°C)	Results mode	T_i (μs)	U_i (kV)	L_{fi} (m)	t_b (μs)
Hemisphere	6	10	600/10 000	1890	1011	21	Exp results	97 ± 25	977 ± 180	6.5	395 ± 28
$\varphi = 100$ mm							Model results	101 ± 4	1001 ± 14	6.44 ± 0.1	396 ± 7
Sphere	8	8	550/10 000	1932	990	21	Exp results	145 ± 36	1398 ± 54	3.62 ± 0.4	354 ± 50
$\varphi = 600$ mm							Model results	147 ± 8	1387 ± 30	3.97 ± 0.42	348 ± 43
Tube	6	6	400/10 000	1445	1019	20	Exp results	140 ± 13	1165 ± 47	3.2 ± 0.22	400 ± 40
$\varphi = 100$ mm							Model results	139 ± 8	1196 ± 13	3.17 ± 0.11	365 ± 12

given experimental configuration, namely T_i, U_i, L_{fj}, and t_b, that were predicted by our model are in good agreement with those obtained experimentally.

Figures 9.3 and 9.4 give some characteristic results of the simulation related to two kinds of gap lengths subjected to the same voltage waveform. The values of these characteristics (leader and return stroke currents, injected charge in the gap, leader radius evolution, power and energy injected in the gap, gradient in the leader, and leader instantaneous velocity) are in good agreement with those obtained experimentally.

Figure 9.5 depicts the experimental and simulated results for the leader current and velocity for a 10 m cone where the plane gap length is subjected to a 500/10 000 μs waveform. Note that the time step for the velocity simulation is the same as the digital sampling interval of the experiment. We observe once more that the simulated results match the experimental ones well, in terms of both amplitudes and waveforms.

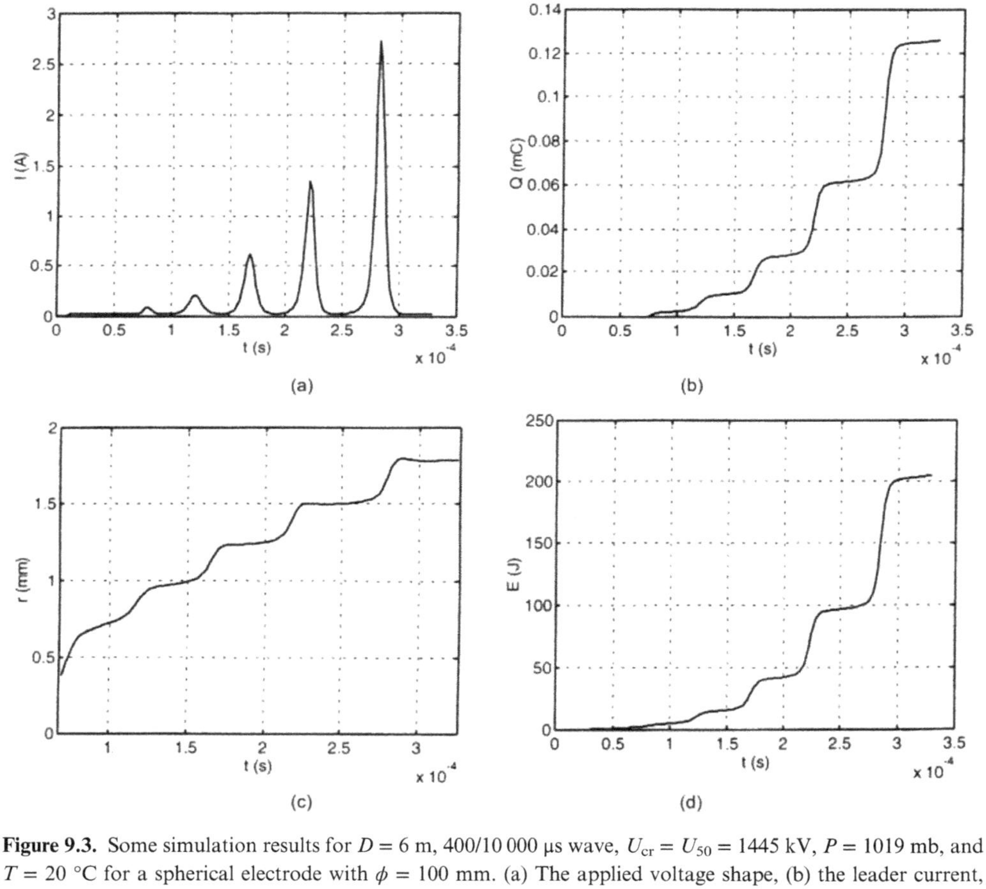

Figure 9.3. Some simulation results for $D = 6$ m, 400/10 000 μs wave, $U_{cr} = U_{50} = 1445$ kV, $P = 1019$ mb, and $T = 20$ °C for a spherical electrode with $\phi = 100$ mm. (a) The applied voltage shape, (b) the leader current, (c) the charge injected into the gap, (d) the evolution of the leader's radius, (e) the leader velocity, (f) the power injected into the gap, (g) the energy injected into the gap, and (h) the gradient in the leader. Reproduced from [12] with permission.

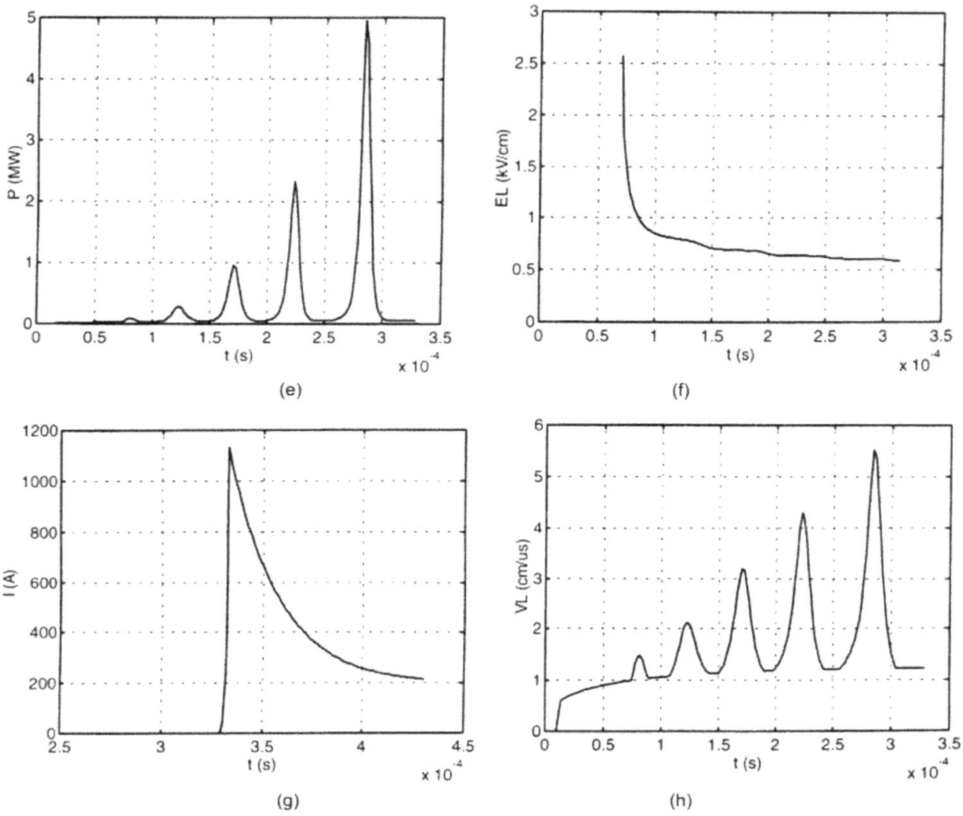

Figure 9.3. (Continued.)

The model adequately describes the evolution of the whole discharge (leader, final jump, and return stroke phases), in contrast to those reported in the most recent literature. Indeed, currently, different models describing the physical processes that occur in each discharge phase are available. However, all these models are limited to single phases and cannot be consolidated in a general model, able to predict the behaviour of a given gap subjected to a given voltage.

It has already been shown that the circuit models have a valuable part to play in the understanding of the plasma and some related processes [2–7]. This model can be presented as a confirmation of this standpoint.

One of the most interesting properties of our model is the fact that for given specific conditions (voltage, gap geometry, some atmospheric conditions, etc), it is possible to determine the leader and the return stroke currents, the corresponding charges, the energy and power injected into the gap, the potential gradient in the leader, the temporal evolution of its thermal radius, the leader propagation velocities and its trajectory, and the simulation of an image converter working in streak or frame mode for the same shot.

The current (or electrical charge) calculated from the model may be readily used to compute radiated electromagnetic fields using analysis presented elsewhere [6–8]. Knowledge of these radiated fields allows us to study their interaction with engineering systems.

9.3 Prediction of the switching impulse withstand voltages of long air gaps

The determination of the critical breakdown voltage constitutes an important practice parameter for the design of HV engineering systems and especially for outdoor insulation where air is used as the insulation medium [9, 10]. Generally, this important parameter is determined experimentally after long and often costly tests. For example, the testing program on a 765 kV line [9], an extremely expensive and time-consuming project, shows that the determination of the required clearances of

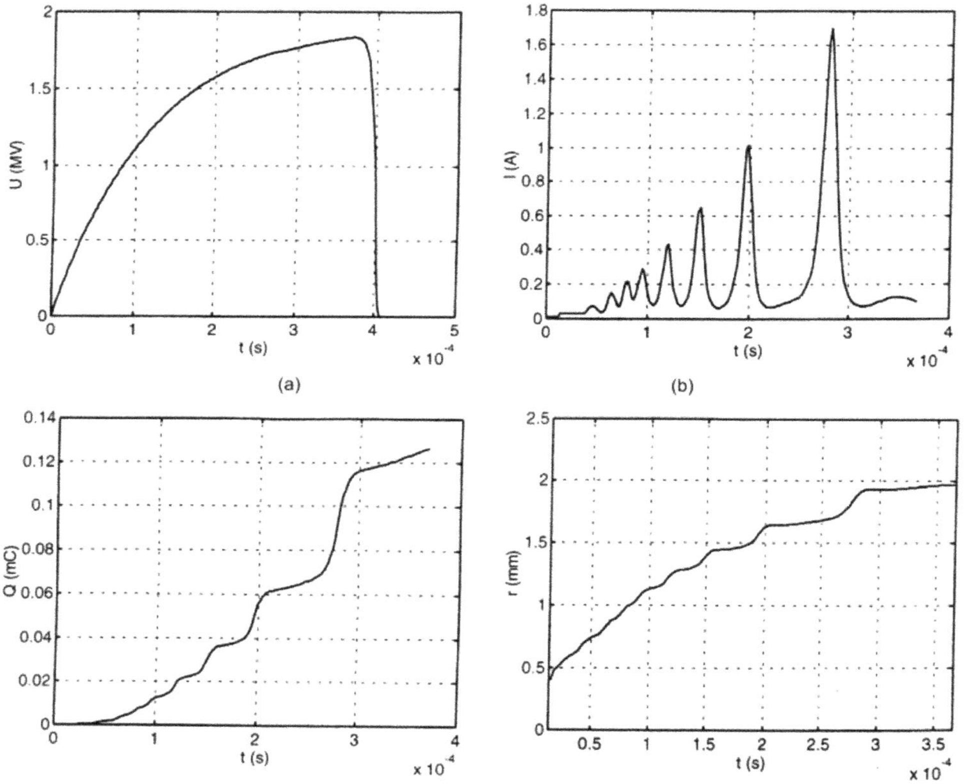

Figure 9.4. Some simulation results for $D = 8$ m, 550/10 000 μs wave, $U_{cr} = U_{50} = 1932$ kV, $P = 990$ mb, and $T = 21$ °C for a spherical electrode with $\phi = 600$ mm. (a) The applied voltage shape, (b) the leader current, (c) the charge injected into the gap, (d) the evolution of the leader's radius, (e) the leader velocity, (f) the power injected into the gap, (g) the energy injected into the gap, and (h) the gradient in the leader. Reproduced from [12] with permission.

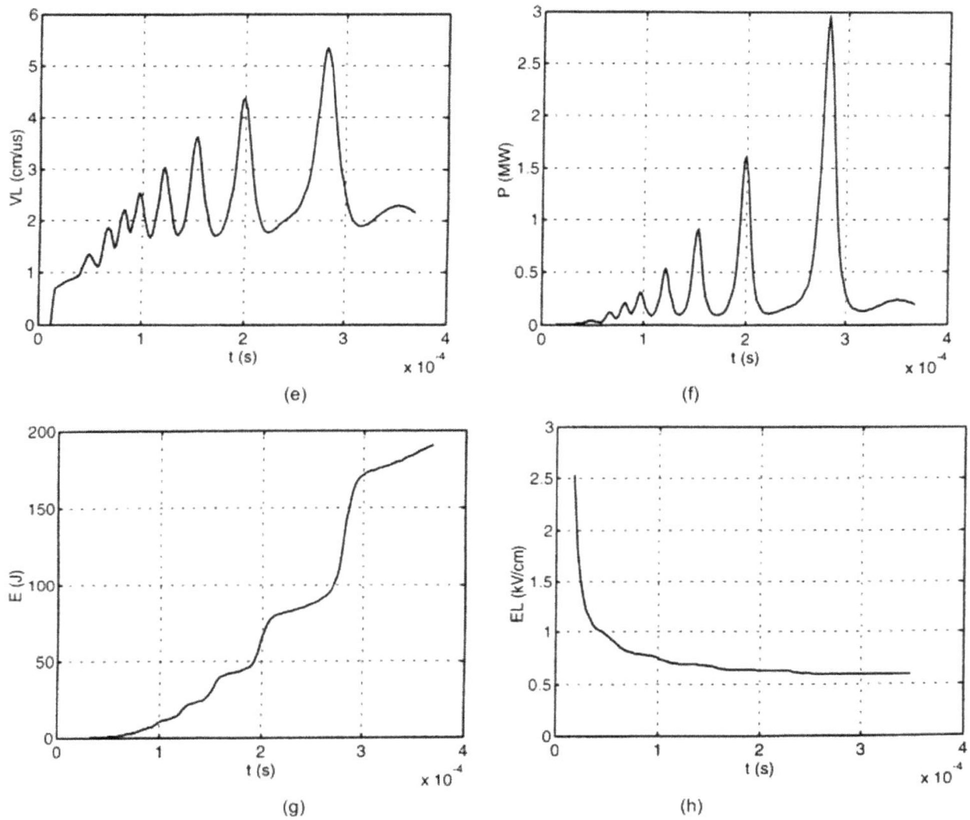

Figure 9.4. (Continued.)

transmission line design required a full-scale test. Mathematical models that simulate discharge processes in long air gaps might be useful for the prediction of breakdown characteristics over a wide range of geometries and atmospheric conditions; therefore reducing the required full-scale testing.

Research on the air gap withstand has therefore been primarily directed towards the understanding of the growth of the discharge and to develop predictive mathematical models [11]. The aim of these models was to provide numerical tools to predict, as accurately as possible, the critical flashover voltage (CFOV) and to simulate the development of electric discharges. A good theoretical model is an important asset as it may be used as a preliminary design tool and help in minimizing experimental efforts.

This section is aimed at determining the U_{50} breakdown voltages of long air gaps subjected to positive/negative switching impulse voltages using predictive dynamic models previously developed by the authors [4, 12–14]. These latter are self-consistent models based on an equivalent electrical network; the parameters of which vary with time according to the discharge characteristics and geometry. The salient feature of the model is that it takes into account the gap geometry while including atmospheric

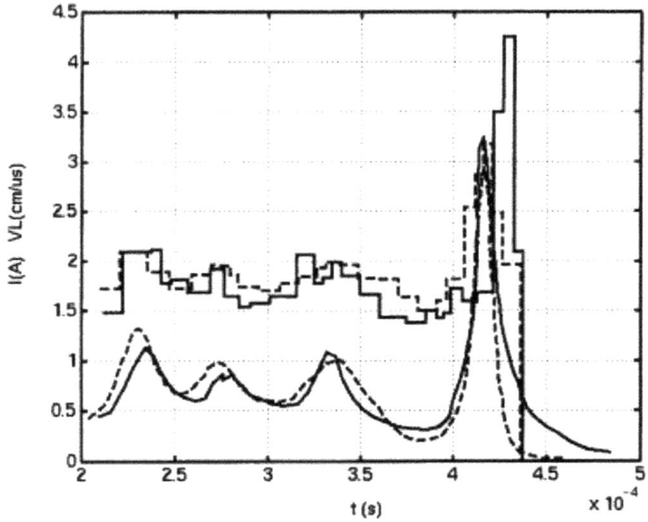

Figure 9.5. The variation of the leader current and velocity for D = 10 m, a cone–plane gap, 500/10 000 μs wave, $U_{cr} = U_{50}$ = 1890 kV, P = 1011 mb, and T = 21 °C. The full line represents the experimental values [1] and the dashed line shows the simulated values. Reproduced from [12] with permission.

correction factors. The input data are the geometry of electrodes, the voltage wave shape, and the ambient air pressure and temperature.

In addition to the computation of some major parameters, including time lag to breakdown, discharge current, and propagation velocity which describe the temporal discharge propagation up to breakdown [4, 12–14], the model allows the prediction of the breakdown voltage. The computed results correlate well with experimental data under various conditions/configurations.

The breakdown mechanisms of the long air gap can be explained by the classic theories of corona-leader mechanisms [1, 14]. A large number of papers and books on the behaviour of corona-streamers are available [1, 15–20]. Corona inception is known to be a statistical event [21–23]. Basically, three conditions must be met to obtain streamer inception: (i) an initial electron must be available, (ii) the applied field need to be above the so-called critical field strength, and (iii) the needs to be a sufficient or critical volume characterized by a critical distance.

The breakdown in air depends strongly on the gap configuration, wave shape, and polarity of the surge as well as the ambient conditions. An overview of these factors and the way they affect the flashover voltage can be found in the literature [18]. The strength of the insulation to positive polarity switching surges is widely used to estimate the required clearances in transmission line tower windows.

The gap factor, k, is defined as the ratio of the CFVO of any other configuration to the CFOV of a rod–plane gap, tested with the same gap spacing and wave shape and with a positive polarity switching impulse. This gap factor, being independent of gap spacing, wave shape, and weather conditions, is very useful for describing the effects of gap geometry on flashover voltage. This gap factor also becomes an extremely powerful tool for predicting the flashover voltage of a given configuration to a positive polarity switching impulse.

To estimate the minimum CFOV, one uses the relationship (9.1) proposed by Gallet [24] that is valid for rod–plane air gaps D ranging from 6–12 m,

$$U_{50} = \frac{3400}{1 + \frac{8}{D}}. \tag{9.1}$$

Typical gap factors used for station clearances can be found in [18]. Since the gap factor for a specific configuration is unknown, a value of 1.30 is suggested.

Several papers have been dedicated to studying the influence of weather conditions [10, 18, 23, 25]. The rules for the atmospheric correction of withstand voltages for external insulation are specified in the IEEE Standard [10]. The effects of air density and humidity are the most prominent weather conditions affecting the flashover voltage; the flashover voltage of an air gap mainly varies with the gap spacing and wave shape of the voltage [25]. As the gap spacing increases, the flashover voltage becomes less sensitive to changes in weather conditions [9, 10, 18] and the appropriate weather corrections diminish. For long air gaps, the weather correction becomes more dramatic as the wave front increases. Generally, as the relative air density increases, the flashover voltage also increases, similar to the effect of increasing the gap spacing. As the time to crest increases, the CFOV decreases until it reaches a minimum that corresponds to the critical time to crest and then increases, forming a U-shaped curve [18].

The relationship between the positive polarity flashover voltage and the gap spacing is non-linear [18]. As the gap spacing increases, the flashover voltage becomes less sensitive to changes in gap spacing.

$$U_{50} = 1118D^{0.45} \tag{9.2}$$

Experimental results reported by the Les Renardières Group [1] reveal the general tendencies of the variations in breakdown voltage versus the radius of a HV electrode. For long air gaps, it appears that the radius of curvature of a HV electrode point has an insignificant contribution to the value of U_{50}, especially with voltages with short and long rising times.

Contrary to the positive polarity for which there is a critical radius above which U_{50} is influenced, it seems that there is not such a radius for a negative polarity. By comparing the variation of U_{50} for different radii of curvature, Pigini *et al* [17] observed that U_{50} is more sensitive to voltage variation under positive than negative surges.

The humidity (g m^{-3}) of air influences the development of long sparks. By carrying out some experiments over several days on an air gap of 8 m for humidity varying between 8 and 17 g cm^{-3}, it has been observed that the U_{50} voltage increases by 10% [18]. The dependency is more pronounced for long air gaps ($D > 5$ m) and high humidity ($H > 10$ g m^{-3}); the humidity of air under normal conditions is 11 g m^{-3}. This value is due to the electronegative character of humidity which is observed at a highly divergent electric field.

9.3.1 Prediction of positive switching impulse withstand voltages

To validate the proposed model, some parameters derived from our predictive models [5, 12] were compared to experimental results reported by the Les Renardières Group [1]. The overall results are compared with an experiment that has the same number of shots and the same conditions (electrode shapes, atmospheric conditions, and applied voltage waveform) and important parameters, such as the corona inception time and voltage, the final jump height, and time to breakdown, are determined. In each case, the crest voltage corresponds to U_{50} (figures 9.6–9.9).

As is evident from these figures, all the parameters characterizing the discharge, namely the corona inception time and voltage, the final jump height, and time to breakdown, are predicted well by the model, compared with those obtained experimentally. However, the standard deviation predicted by the model is generally low compared to those recorded experimentally. The erratic propagation of the leader channel and the statistics of time to breakdown related to the initiation of the first corona are important parameters behind the statistical dispersion (scattering) of breakdown processes [1, 10, 14, 18]. Another important parameter not taken into account in the model, i.e. the influence of humidity, might be behind this difference. Water vapour present in air increases electronegativity, especially under a divergent electric field [26].

9.3.2 Prediction of the 50% positive breakdown voltage U_{50}

To design air gap clearances, it is necessary to assess the breakdown probability of the various air gaps and components of recovery-type insulation systems. The strength of insulation under a switching impulse is a statistical parameter linked with

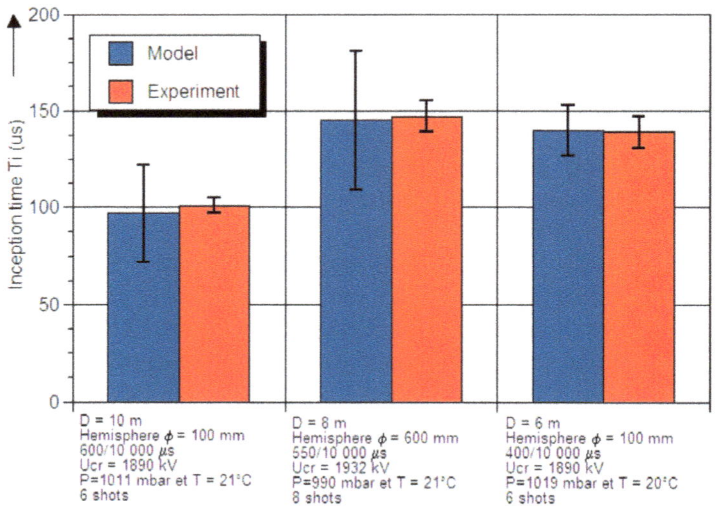

Figure 9.6. Comparison of the corona inception time computed from the model with results obtained experimentally by the Les Renardières Group [7], under the same conditions for positive polarity. © 2013 IEEE. Reprinted, with permission, from IEEE Transactions on Dielectrics and Electrical Insulation.

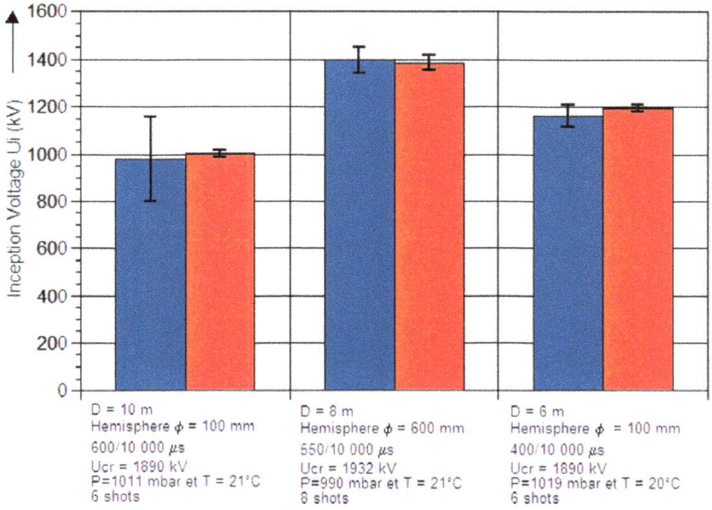

Figure 9.7. Comparison of the corona inception voltage computed from the model with results obtained experimentally by the Les Renardières Group [7], under the same conditions for positive polarity. © 2013 IEEE. Reprinted, with permission, from IEEE Transactions on Dielectrics and Electrical Insulation.

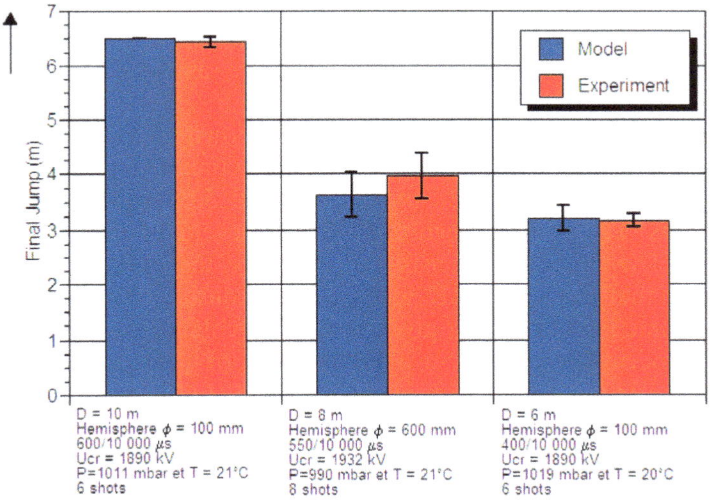

Figure 9.8. Comparison of the final jump height computed from the model with results obtained experimentally by the Les Renardières Group [7], under the same conditions for positive polarity. © 2013 IEEE. Reprinted, with permission, from IEEE Transactions on Dielectrics and Electrical Insulation.

a probability of flashover that varies with the switching impulse magnitude polarity and wave shape. The probability of a switching impulse flashover can be modelled with a normal curve, with the CFOV defined as the mean. The CFOV is the switching impulse voltage level at which the probability of a flashover is 50%.

The 'up-and-down' method is widely used for estimation of the 50% probability breakdown voltage U_{50}. For most breakdown probability functions, the scaling

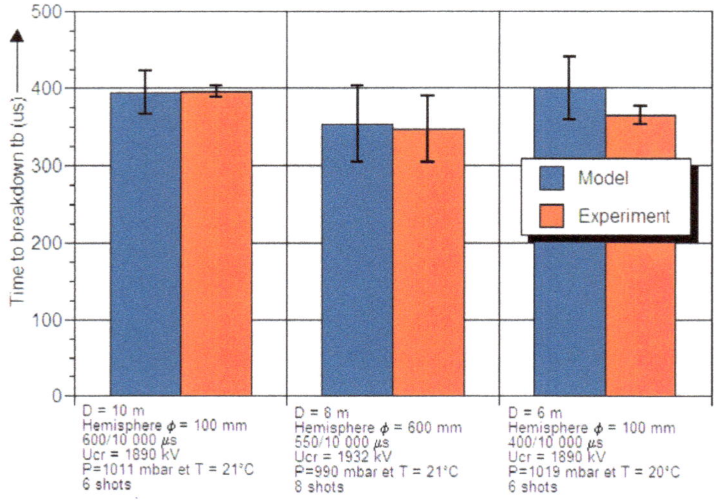

Figure 9.9. Comparison of the time to breakdown computed from the model with results obtained experimentally by the Les Renardières Group [7], under the same conditions for positive polarity. © 2013 IEEE. Reprinted, with permission, from IEEE Transactions on Dielectrics and Electrical Insulation.

parameter is simply the standard deviation σ. The classical up-and-down method of Dixon and Mood is based on a maximum-likelihood estimation of U_{50} and σ for a normal distribution [20]. The CFOV was, therefore, assessed by means of the up-and-down method.

The determination of the CFOV or U_{50} breakdown voltage of a given air gap requires a statistical investigation. The U_{50} voltage represents the critical value of the crest voltage U_{cr} leading to 50% breakdown and 50% withstand from a given number of shots. The up-and-down method was developed to estimate the U_{50} or CFOV value. The procedure consists in applying the voltage at various levels A_i selected such that $U_{i,1} = U_i + \mathrm{d}U$, where $\mathrm{d}U$ is a constant voltage increment. Only one shot at a time is applied at a given level. The voltage level is then changed to U_{i-1}, if the application resulted in a breakdown, or to U_{i+1} if the result was a 'withstand'. The test sequence considered in these studies consisted of a total of $n = 50$ shots (voltage applications), which is considered to be accurate [11, 18].

A normal probability distribution is assumed. By using the up-and-down method, and by performing a series of n shots, the exact maximum-likelihood estimates of U_{50} voltage are obtained using the following relationship:

$$U_{50} = \frac{1}{n}\sum_{k=1}^{n} U_{cr}(i). \tag{9.3}$$

Knowledge of U_{50} and the standard deviation σ permits us to define the U_0 (theoretical voltage level, below which no breakdown is possible) as

$$U_0 = U_{50}(1 - 2.5\sigma). \tag{9.4}$$

At the same time, the voltage U_{100} above which the gap will always breakdown is

$$U_0 = U_{50}(1 + 2.5\sigma). \tag{9.5}$$

For a given electrode configuration and atmospheric conditions subjected to a given voltage, the model allows us to determine parameters of practical interest such as the

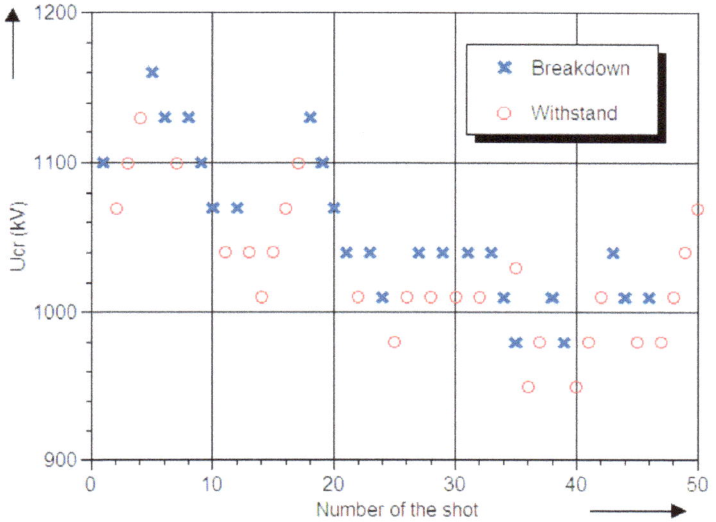

Figure 9.10. Simulation of an up-and-down sequence: D = 2.6 m, –320/10 000 μs wave. × is breakdown and ○ is withstand. U_{50} = 1040.6 kV and σ = 4.8%. © 2013 IEEE. Reprinted, with permission, from IEEE Transactions on Dielectrics and Electrical Insulation.

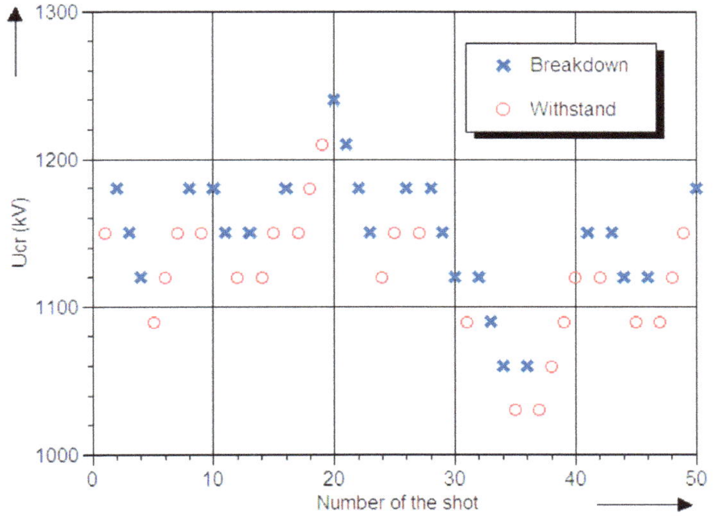

Figure 9.11. Simulation of an up-and-down sequence: D = 3 m, –320/10 000 μs wave. × is breakdown and ○ is withstand. U_{50} = 1135 kV and σ = 3.89%. © 2013 IEEE. Reprinted, with permission, from IEEE Transactions on Dielectrics and Electrical Insulation.

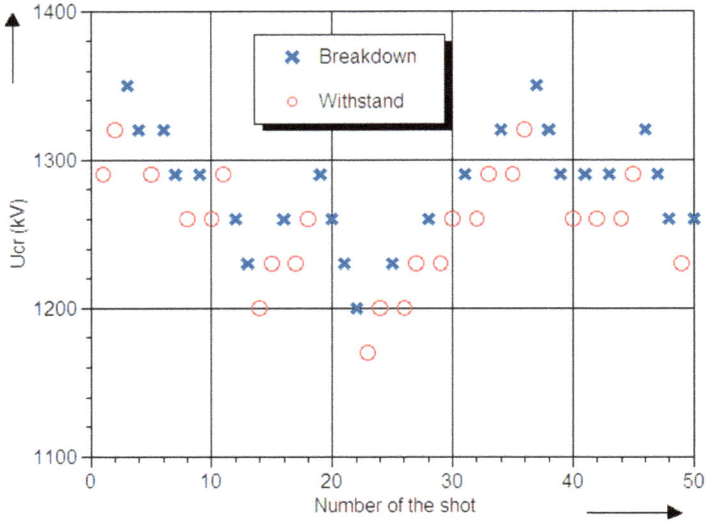

Figure 9.12. Simulation of an up-and-down sequence: D = 4 m, –320/10 000 μs wave. (×) breakdown and (○) withstand. U_{50} = 1269 kV and σ = 3.1%.

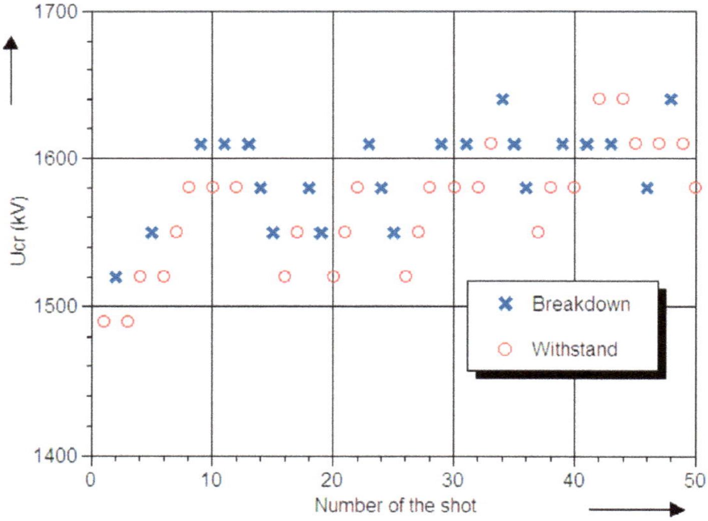

Figure 9.13. Simulation of an up-and-down sequence: D = 7 m, –320/10 000 μs wave. × is breakdown and ○ is withstand. U_{50} = 1577.6 kV and σ = 2.4%. © 2013 IEEE. Reprinted, with permission, from IEEE Transactions on Dielectrics and Electrical Insulation.

50% breakdown voltage or the time to breakdown. The computed values of CFOV have been compared with experimental results obtained on rod–plane gaps ranging from 2–10 m (a hemispherical electrode tip with a 0.6 m diameter) and subjected to a positive switching impulse of 320/10 000 μs [1]. Figures 9.10–9.14 show the

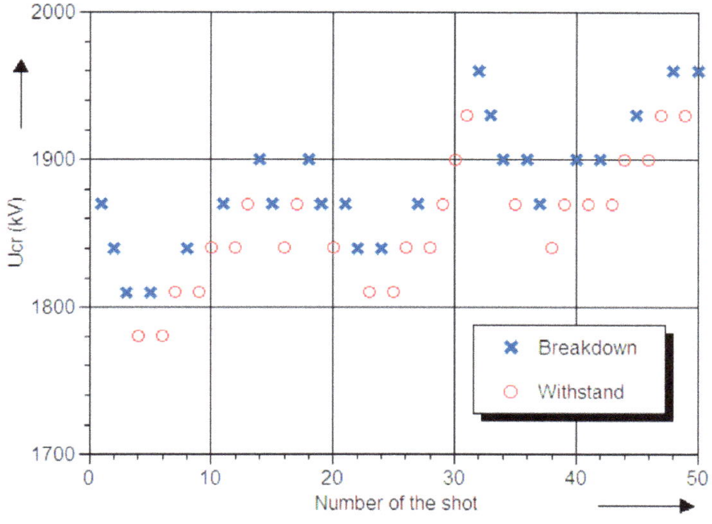

Figure 9.14. Simulation of an up-and-down sequence: $D = 10$ m, –320/10 000 μs wave. × is breakdown and ○ is withstand. $U_{50} = 1869$ kV and $\sigma = 2.4\%$. © 2013 IEEE. Reprinted, with permission, from IEEE Transactions on Dielectrics and Electrical Insulation.

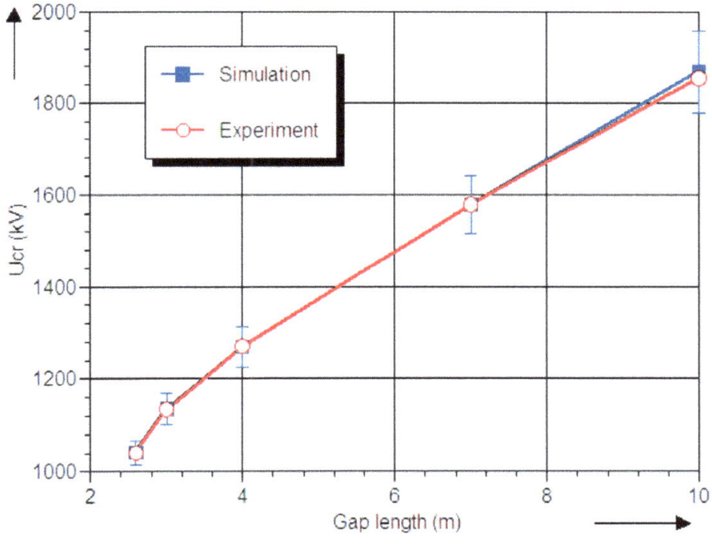

Figure 9.15. Comparative summary of the U_{50} simulated results with experimental data reported by the Les Renardières Group [1]. The U_{50} voltage was determined as a function of the gap length D. © 2013 IEEE. Reprinted, with permission, from IEEE Transactions on Dielectrics and Electrical Insulation.

simulations of the CFOV voltage using the up-and-down method with 50 shots for each case.

The standard deviation (σ) is between 2.4% and 4.8%. These values are consistent with experimental values, where the dispersions vary between 4% and 5% [1]. The results obtained for CFOV or U_{50} are thus compared to experimental results. Figure 9.15 compares the theoretical and experimental CFOV as a function of the gap length D. Good agreement can be observed between these results.

9.4 Flashover voltage of long air gaps in the presence of a floating insulating barrier

The barriers effect was unveiled by E Marx and H Rozer while studying discharge in air gaps [27]. The importance of this phenomenon lies in the fact that the insertion of a thin solid insulator (barrier) in a point–plane gap in a gas (in air in particular) or a dielectric liquid, increases both the breakdown voltage of this gap and the time to breakdown [27–30]. Such an increase is linked to the field redistribution in the insulation gap after the initiation of the discharge and/or the elongation of the minimum leakage length. It is agreed that such a field redistribution is due to the excess of charge deposition on the barrier induced by the electrode injection or ionization processes.

The barrier effect depends upon many parameters, the most important of which are the electrode configuration (i.e. their geometry and nature), the position and dimensions of the barrier with respect to the electrodes, the polarity and type (waveform) of the applied voltage, and the surface conductivity of the barrier. Some studies [31] have shown that the effect of a barrier on the dielectric strength of point–plane air gaps is more significant for a positive point than for a negative point. The effect can even be detrimental when the point is negative, especially if the barrier is near the plane electrode. The optimal position of the barrier for which the breakdown voltage reaches its maximum value is generally close to the vicinity of the HV electrode but not touching it [27–31].

The aim of this section is to show the utility of the established model in another insulating structure, namely a point–plane electrode arrangement in which an insulating barrier is inserted. Thus, we investigate theoretically as well as experimentally the influence of insulating barriers on the dielectric strength of point–plane air gaps under positive lightning and switching impulses.

9.4.1 Modelling the barrier effect

It has already been pointed out that the increase in the dielectric strength through the insertion of a floating barrier is mainly due to the fact that the barrier acts as a geometric obstacle to direct discharge. The discharge grows from the point (HV) to the barrier's edge and from this edge to the grounded plane. Since the aim of a modelling is to attain the simplest model that properly describes the experimental observations, we use the following simple analytical method (figure 9.16) instead of calculations using classical methods, such as equivalent charges, finite element methods, and the Monte Carlo method, all of which require long computation times.

The discharge propagation axis is recalculated at each step of the propagation through the angle θ_i (figure 9.17) as

$$tg\theta_i = \frac{L - x_i}{a_1 - y_i} \quad \text{for} \quad y_i \leqslant a_1 + a_2 \tag{9.6}$$

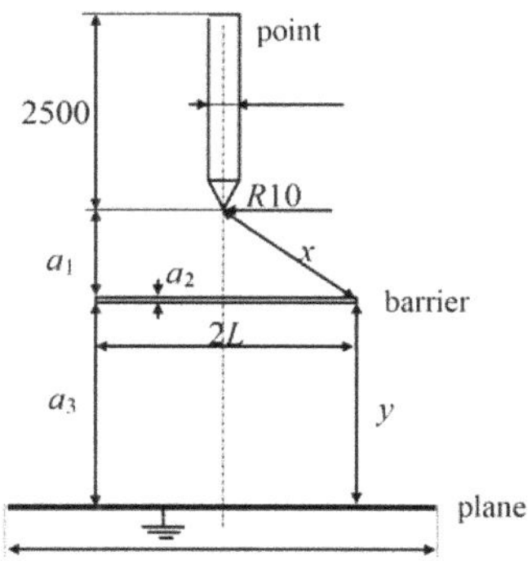

Figure 9.16. Arrangement of electrodes and barrier.

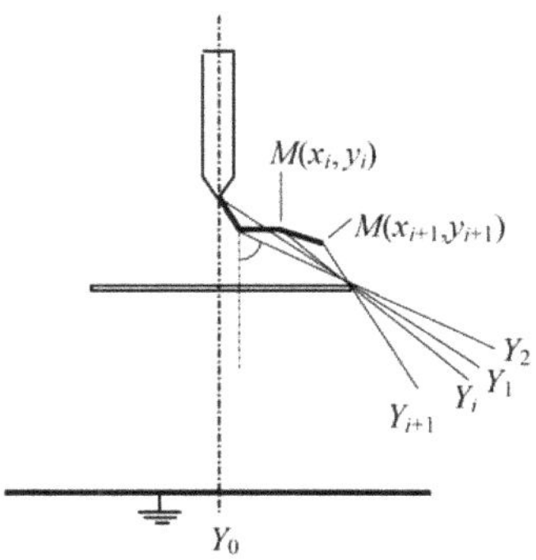

Figure 9.17. The simple assumption used to model the barrier effect.

or

$$\theta_i = 0 \text{ for } \quad y_i \geqslant a_1 + a_2. \tag{9.7}$$

9.4.2 The gap capacitance

The electrode gap is divided into three partials gaps. The electric field E_1 in a_1 remains non-uniform, but in a_3 it becomes uniform and equal to E_3. The field E_2 in a_2 is also uniform. Such assumptions have been experimentally verified by Wolochtchenko [2] using a dielectric probe. The capacitance C_0 of the electrode gap is, therefore, the sum of three capacitances as follows:

$$C_0 = \left[\frac{1}{C_{a1}} + \frac{1}{C_{a2}} + \frac{1}{C_{a3}}\right]^{-1}. \tag{9.8}$$

C_{a1} is obtained by using spherical approximation [32]:

$$C_{a1} = 4\pi\varepsilon_0 R_p \left(\frac{a_1}{a_1 - R_p}\right). \tag{9.9}$$

C_{a2} and C_{a3} are plane capacitances:

$$C_{a2} = \varepsilon_r \frac{b^2}{a_2} \tag{9.10}$$

and

$$C_{a3} = \varepsilon_0 \frac{b^2}{a_3}. \tag{9.11}$$

In the following, we present and discuss the results of an experimental and theoretical modelling study concerning the influence of a clean barrier on the breakdown voltage for a point–plane air gap under lightning and switching impulse voltages. The polarity of the point is considered exclusively as being positive.

9.4.3 Experiment

The HV electrode consists of a steel needle point 10 mm in radius (conical in shape 30°). The grounded plane electrode is a circular steel plate 4 m in diameter, 1.5 mm thick. The bakelized paper barriers ($\varepsilon_r = 5.8$) are squares with 1.3 m sides and 6 mm thick. This thickness is sufficient to prevent the electrical breakdown of the barrier and to ensure good mechanical resistance. The barrier is suspended by insulating supports (four nylon filaments 4 mm in diameter). The electrode gap D is about 1.5 m. The arrangement of the electrode and barrier configuration is shown in figure 9.15. The lightning and switching impulse voltages are supplied by a Marx generator rated at 2800 kV, 32 kJ.

9.4.4 Results and discussion

The breakdown voltage is determined by the up-and-down method on a series of 50 impulses using the experimental simulator [33] as practiced experimentally in the laboratory.

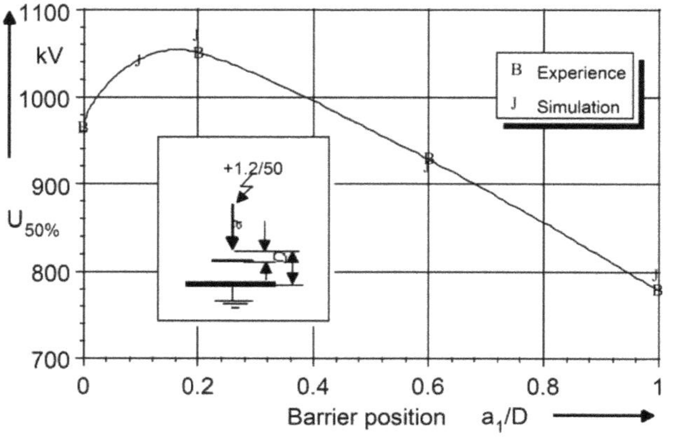

Figure 9.18. Experimental and simulated results of U_{50} breakdown lightning impulse voltage with respect to the barrier position a_1.

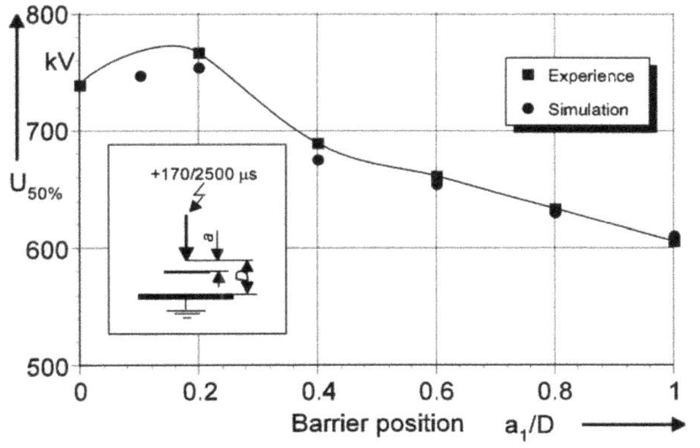

Figure 9.19. Experimental and simulated results of U_{50} breakdown switching impulse voltage with respect to the barrier position a_1.

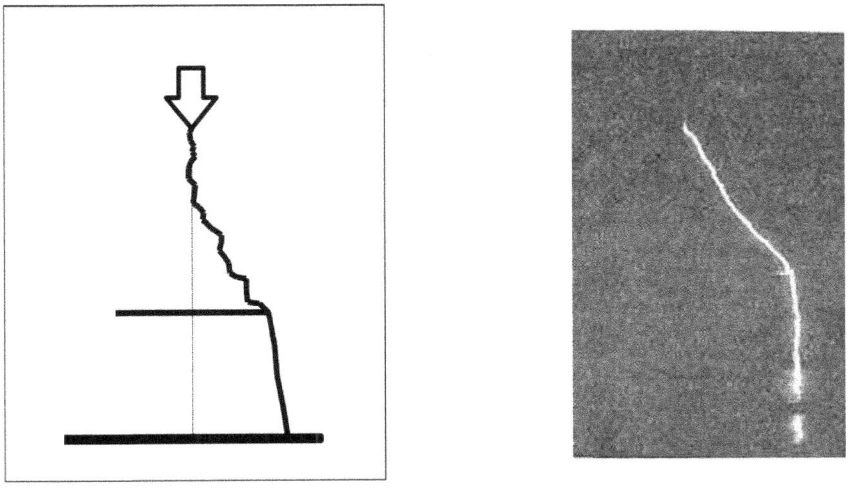

Figure 9.20. Discharge trajectory. (a) Simulation of the discharge trajectory as we were able to photograph it, with voltage $U_{cr} = 920$ kV, $D = 1.5$ m, and $a_1 = 0.9$ m. (b) Photograph of the discharge trajectory, with voltage $U_{cr} = 920$ kV, $D = 1.5$ m, and $a_1 = 0.9$ m.

For the experimental investigations, the interval between two successive impulses is 60 s. This interval is sufficient to eliminate any influence of the surface charge on the statistical determination of the U_{50} breakdown voltage. Indeed, earlier work was performed by varying this interval from 30–300 s, and no differences in the U_{50} breakdown voltage for $t > 60$ s were observed [34]. Note that the simulations were not performed for barriers near the HV electrode. In this case, superficial sliding discharges appearing on the surface of the barrier facing the point are not taken into account by our model.

Figures 9.18 and 9.19 give the results of the theoretical and experimental investigations. We observe that the theoretical results are consistent with the trials, and the optimum position of the barrier is found to be about $0.2D$, i.e. at 20% of the point–plane gap.

Figure 9.20 shows a photograph and a computer simulation of the discharge under the same experimental conditions, i.e. for a lightning impulse voltage with a peak value of $U_{cr} = 920$ kV, $D = 1.5$ m, and $a_1 = 0.9$ m. The theoretical results once again are comparable to experimental results.

References

[1] Les Renardieres Group 1977 *Electra* **53** 31

[2] Fofana I and Beroual A 1995 *J. Phys. D: Appl. Phys.* **28** 303

[3] Fofana I and Beroual A 1995 Modelling of the leader and the return stroke currents with an equivalent electrical network *9th Int. Symp. High Voltage Engineering* (*Graz, Austria, August* 28–*September 1*) paper 2141

[4] Fofana I and Beroual A 1996 *IEEE Trans. Dielectr. Electr. Insul.* **3** 273

[5] Ratnamahilan P 1993 *IEEE Trans. Magn.* **29** 1839

[6] Taudière I 1988 Analyse du rayonnement électromagnétique VHF-UHF de l'éclair—application à un modèle d'avancée du précurseur négatif *PhD Thesis* Université Paris VI

[7] Labaune G, Richard P and Bondiou A 1987 *Lighting Electromagnetics* (New York: Hemisphere) p 261

[8] Rathoin S 1993 Contribution à la modélisation du rayonnement électromagnétique de la foudre en vue du couplage sur les câbles *PhD Thesis* Ecole Centrale de Lyon, University of Lyon

[9] Hill K H 1999 Estimation of positive polarity switching surge flashover voltage of compact transmission lines *MSc Thesis* University of Manitoba, Winnipeg

[10] IEEE 1995 *Standard Techniques for High Voltage Testing* IEEE Std 4-1995

[11] Working Group 07 (Dielectric strength of external insulation under transient voltage) of Study Committee 33 (Over-voltages and Insulation Coordination) 1992 Guidelines for the evaluation of the dielectric strength of external insulation *CIGRE Brochure* no. 72

[12] Fofana I and Beroual A 1997 *J. Phys. D: Appl. Phys.* **30** 1653

[13] Rakotonandrasana J H, Beroual A and Fofana I 2008 *J. Phys. D: Appl. Phys.* **41** 105210

[14] Beroual A 2004 *REE Rev. Electr. Electr.* **8** 36

[15] Les Renardières Group 1981 *Electra* **74** 67

[16] Hutzler B and Hutzler D 1992 Modélisation de l'amorçage des grands intervalles d'air *Bull. Dir. Etud. Rech.* **4** 11

[17] Pigini A, Rizzi G, Garbagnati E, Porrino A, Baldo G and Pesavento G 1989 *IEEE Trans. Power Deliv.* **4** 1379

[18] Gallimberti I 1979 *J. Physique* **40** 193

[19] Le Roy G, Gary C, Hutzler B, Lalot J and Dubanton C 1984 *Les Propriétés diélectriques de l'Air et les Très Hautes Tensions* (*Collection de la Direction des Études et Recherches d'Électricité de France vol 51*) (Paris: Eyrolles)

[20] Rees T 1997 Contribution à l'étude expérimentale et théorique des décharge électriques de polarité négative dans l'air à pression atmosphérique *PhD Thesis* Université de Pau et des Pays de l'Adour, France

[21] Ortéga P 1992 Comportement diélectrique des grands intervalles d'air soumis à des ondes de tension de polarité positive ou négative *PhD Thesis* Université de Pau et des Pays de l'Adour, France

[22] Craggs J D and Meek J M 1978 *Electrical Breakdown of Gases* (New York: Wiley)

[23] Bazelyan E M and Raizer Y P 1997 *Spark Discharge* (Boca Raton, FL: CRC)

[24] Gallet G, LeRoy G, Lacey R and Kromer I 1975 *IEEE Power Appar. Syst.* **6** 1989

[25] EPRI 1982 *Transmission Line Reference Book: 345 kV and Above* 2nd edn (Palo Alto, CA: EPRI)

[26] Fofana I and Beroual A 1996 *J. Phys. D: Appl. Phys.* **29** 691

[27] Gefle O S and Lebeder S M 1995 Barrier effect in solid dielectrics *9th Int. Symp. on High Voltage Engineering* (*Graz, Austria*) paper 10.70

[28] Beroual A and Boubakeur A 1991 *IEEE Trans. on Electr. Insul.* **26** 1130

[29] Zouaghi A and Beroual A 1997 Barrier effect on the time lag to breakdown of transformer oil gaps under lightning impulse voltage *Conf. on Electrical Insulation and Dielectric Phenomena* (*Atlanta, GA, USA, 17–20 October*) p 640

[30] Beroual A and Zouaghi A 1999 *Arch. Electr. Eng.* **3** 305

[31] Darveniza M and Holcombe B 1977 The switching impulse strength of a long air gap with a plastic sheet barrier *Congrès d'Electrotechnique Mondial* (*Moscow, Russia*) p 104

[32] Fofana I and Beroual A 1997 *J. Phys. D: Appl. Phys.* **30** 1653

[33] Fofana I and Beroual A 1997 Determination of the U_{50} voltage for long air gaps by an experimental simulator *10th Int. Symp. High Voltage Engineering* (*Montréal, Canada, 25–29 August*) p 21

[34] Boubakeur A 1979 Influence of barriers on the electric strength of mean air gap in point/plane arrangements *PhD Thesis* Politechnika Warszawska, Warsaw

Chapter 10

Applications of the model developed for negative discharge in long air gaps

10.1 Introduction

The aim of this chapter is to show that our model works well. Its validity is verified by comparing the computed results with the experimental results from previous research. For that purpose, we first compare the simulated characteristic parameters of negative long air discharges (namely the spatial and temporal evolution of the whole discharge, the current and the corresponding electrical charge, the power and energy injected into the gap, the potential gradient in the leader channel, the voltage at the tip of the discharge, the temporal evolution of the channel thermal radius, and the instantaneous propagation velocity) given by our model with experimental data related to two point–plane electrode gaps measuring 7 m and 16.7 m. Then, we compare the simulated results of U_{50} for several gaps (2.6, 3, 4, 7, and 10 m) with experimental data. Computations are executed using the scientific library MATLAB.

10.2 Simulation of laboratory experiments

The input data we used to validate our model are identical to those of experiments achieved by the Les Renardières Group [1]. These latter concern two point–plane air gaps measuring 7 m and 16.7 m, where the point electrode has a radius of curvature (r_p) equal to 30 cm, subjected to a 60/3000 μs pure impulse voltage; the crest values of the voltage for U_c are 2.0 MV and 3.0 MV. Figures 10.1 and 10.2 give the simulated results of the spatiotemporal evolution of negative discharges and the corresponding characteristics, namely, the leader current, the charge and power injected into the gap, the instantaneous velocity of the leader, the voltage at the leader tip, the applied voltage, and the leader radius, for two cases of withstanding. We observe that the simulations reproduce the general characteristics observed experimentally.

doi:10.1088/978-0-7503-1236-3ch10

Figures 10.1(c) and 10.2(b) give the spatiotemporal evolution of the space stem, electrode, and space leaders, as well as all other relevant characteristics of discharge. When the upward positive space leader joins the negative electrode leader, a strong re-illumination occurs producing a current pulse, as observed in figures 10.1(e) and 10.2(d) Note that the simplified sketch of the real negative discharge path plotted by the Les Renardières Group [1] (figure 10.1(b)) is similar to that predicted by our model (figure 10.1(c)). Unfortunately, the simplified sketch corresponding to figure 10.2(a) was not available for comparison with that computed in figure 10.2(b).

The model also shows an aspect of randomness in the negative electrode leader trajectory (figures 10.1(d) and 10.2(c)). Indeed, for the same input parameters, one can have different simulated trajectories, as occurred when making several shots in the laboratory under the same experimental conditions. With respect to the simulated currents, waveforms were obtained similar to those observed experimentally [1, 3]. Indeed, the current consists of many pulses (figures 10.1(e) and 10.2(d)). The current pulses correspond to the re-illuminations. The amplitude of these

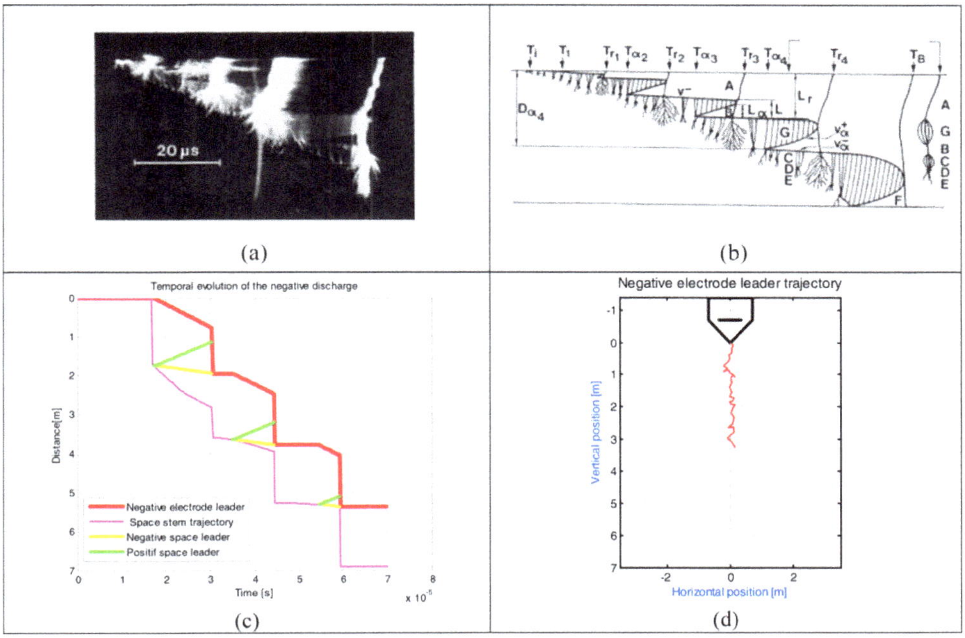

Figure 10.1. A simulation of a point–plane long air gap discharge under a pure impulse shape. Electrode gap $D = 7$ m, applied voltage $U_{cr} = 2.0$ MV, wave $T_1/T_2 = 60/3000$ μs. (a) The real discharge path of a negative discharge for the indicated data in the text. (b) A simplified scheme of the real discharge path of a negative discharge for the data indicated in the text. (c) The simulated results of the temporal evolution of a negative discharge for the data indicated in the text. (d) An example of a simulated leader's trajectory in space. (e) The simulated current of a point–plane long air gap. (f) The simulated charge injected into the electrode gap. (g) The simulated power injected into the electrode gap. (h) Simulated instantaneous velocity. (i) The simulated voltage at the tip of the electrode leader. (j) Simulated applied voltage. (k) Simulated radius of the leader channel. Reproduced with permission from [2].

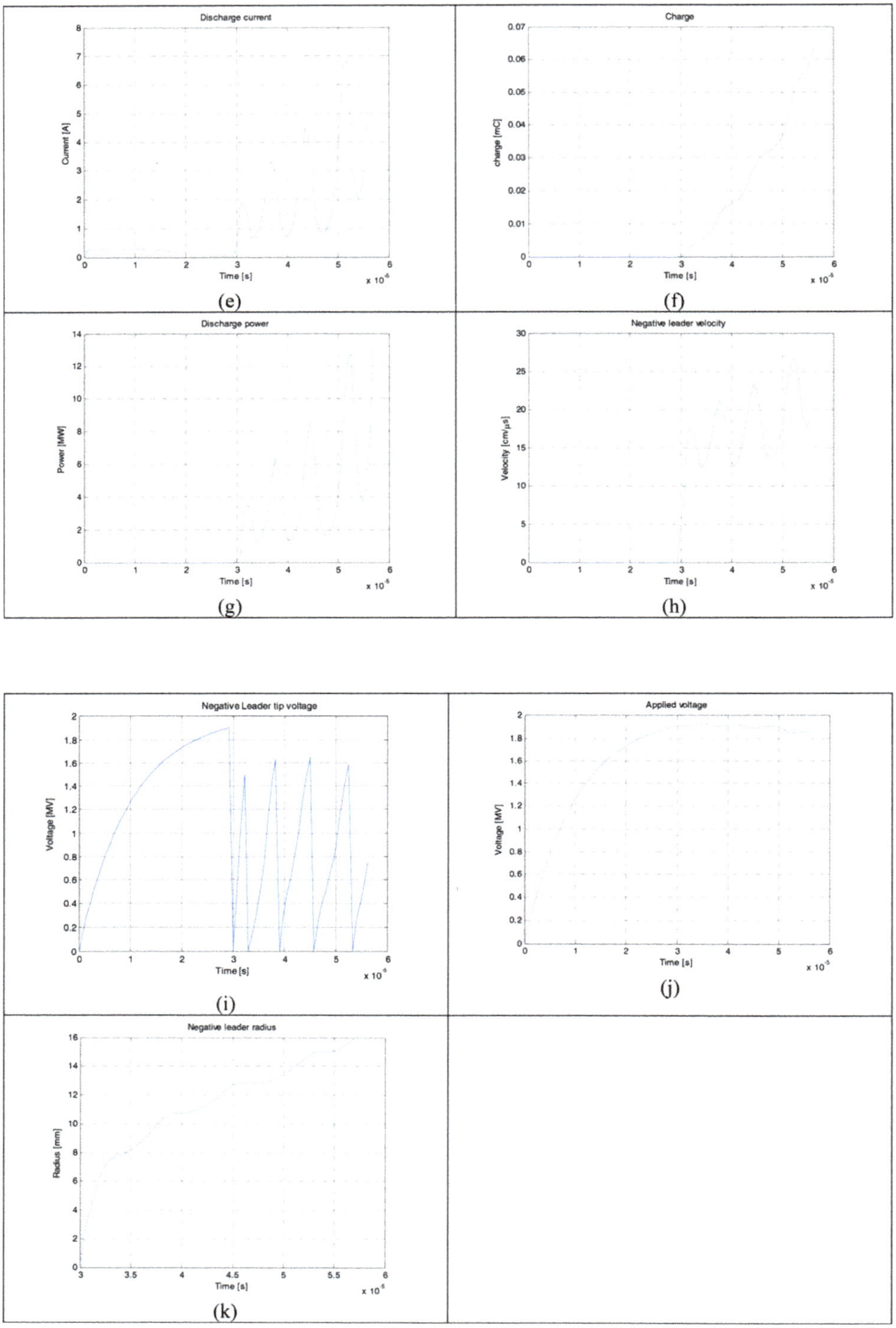

Figure 10.1. (Continued.)

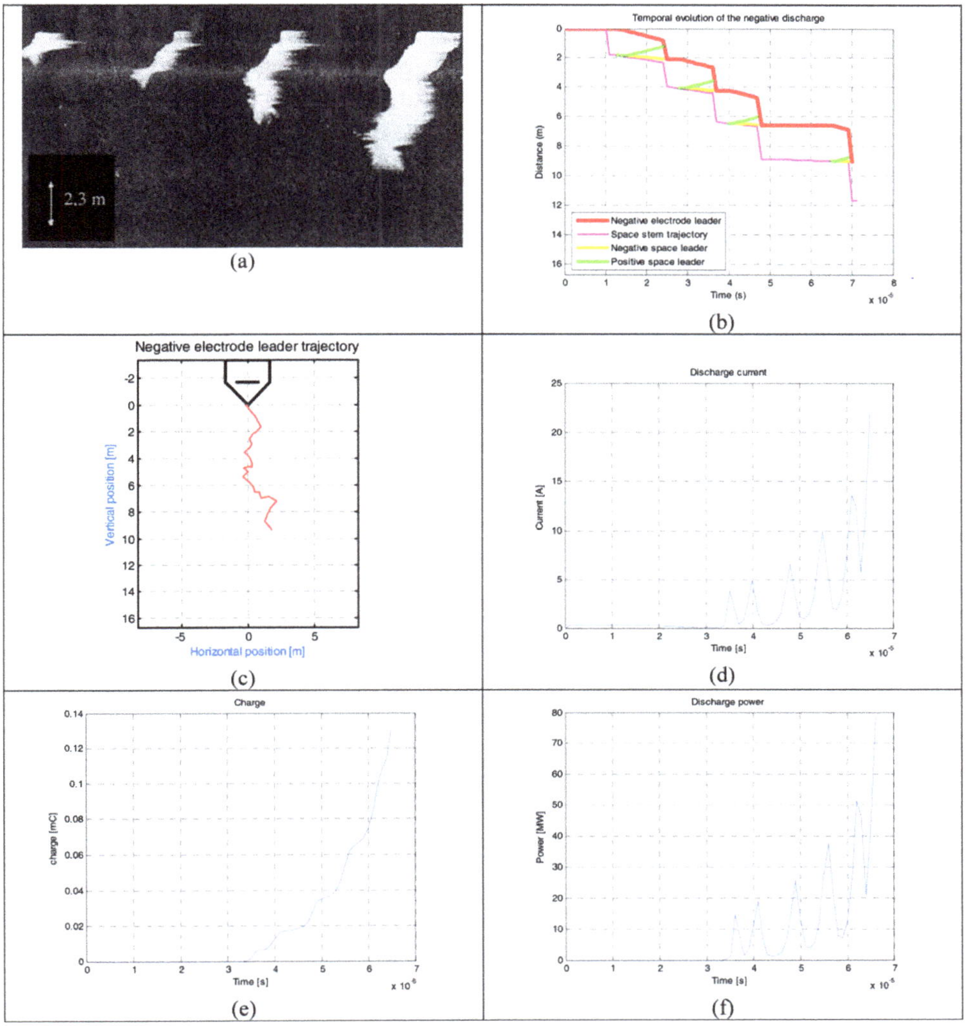

Figure 10.2. A simulation of a point–plane long air gap discharge under a pure impulse shape. Electrode gap $D = 16.7$ m, applied voltage $U_{cr} = 3.0$ MV, wave $T_1/T_2 = 60/3000$ μs. (a) The real discharge path of a negative discharge for the data indicated in the text. (b) Simulated results of the temporal evolution of a negative discharge for the data indicated in the text. (c) An example of a simulated leader's trajectory in space. (d) The simulated current of a point–plane long air gap. (e) The simulated charge injected into the electrode gap. (f) The simulated power injected into the electrode gap. (g) Simulated instantaneous velocity. (h) The simulated voltage at the tip of the electrode leader. (i) Simulated applied voltage. (j) Simulated radius of the leader channel.

current pulses can reach a few amperes with an electrode gap of 7 m—the same range as measured experimentally [1].

However, for the electrode gap of 16.7 m, the amplitudes of these pulses are an order of magnitude lower than those measured experimentally [3]. Indeed, they do

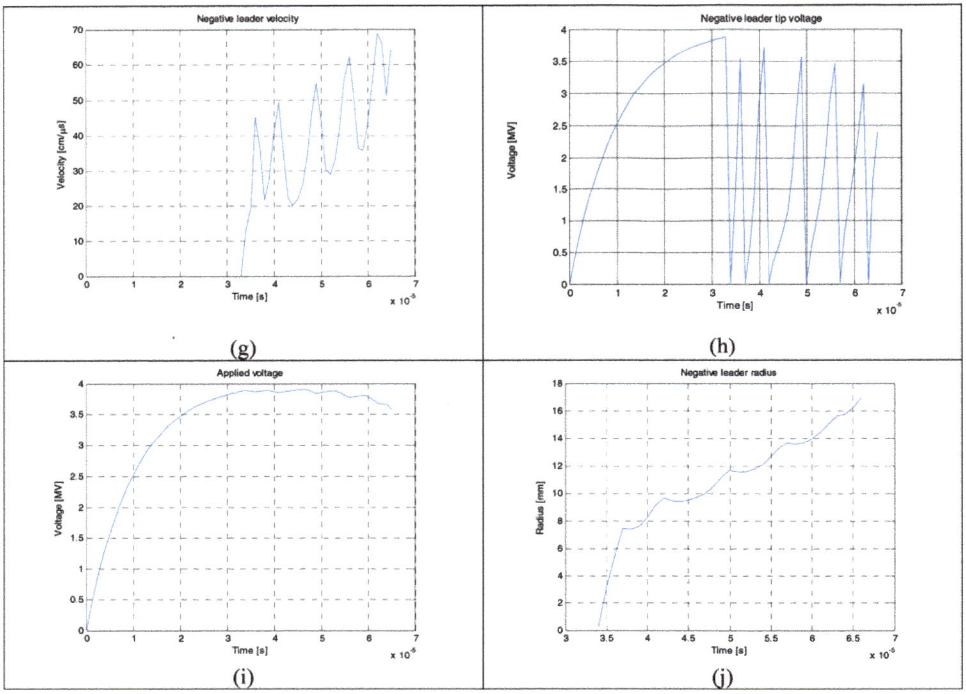

Figure 10.2. (Continued.)

not exceed some tens of amperes while the measured values reached some hundreds of amperes. Note that the current peaks associated with the discharge steps are lower than those produced during the re-illuminations. The number of the latter depends on the applied voltage (waveform and amplitude) and the electrode geometry; hence, they depend on the energy injected into the electrode gap. Moreover, the repetition time of stem occurrence varies between 0.1 and 1 μs, and the time between two re-illuminations lies between 5 and 12 μs (figures 10.1(c) and 10.2(b)).

These results are of the same order as those estimated experimentally [1, 3]. The injected charge is deduced from the discharge current integration (figures 10.1(f) and 10.2(e)). Its variation is found to be step-like. For the experimental conditions, the simulated charge amounts of 60 μC and 130 μC, respectively, for electrode gaps of 7 m and 16.7 m, were considered. The electrical charge is in the same range as that measured for a gap of 7 m, while it is one order dimension lower than that measured for 16.7 m. Such a result is expected since the characteristics of electrical charge depend on those of the current. The radius of negative leader channel varies with the injected charge (figures 10.1(k) and 10.2(j)). The higher the injected charge, the larger the radius is. The initial values of the leader radius are obtained from experimental data [1,3,4].

Knowing the power or the energy injected during the discharge propagation, the negative leader velocity (figures 10.1(h) and 10.2(g)) may be computed. The simulated velocities of the negative discharge can reach 10^5 ms^{-1} which is a value of the same range as that measured experimentally [1, 3]. Note that for a given voltage, the relationship giving the velocity of negative and positive discharges (equation (7.34) or (7.35)) shows that the current (i_L) and velocity (v_L) of the leader are related as $v_L \sim i_L^{1/3}$. Such a relation has been obtained experimentally by Bazelyan *et al* [5], Rizk [6], and Huzler and Huzler for positive discharges [7]. The voltage at the leader tip presents some drops resulting from the re-illuminations phases (figures 10.1(i) and 10.2(h)).

Therefore, the proposed model works well; it constitutes a good tool for the prediction of the characteristics of negative long discharges. The simulated temporal evolution of the negative discharge properly shows the propagation by steps as well as the different sequences of the propagation as reported in the literature. Moreover, the simulated values of the characteristic parameters are generally of the same range as the experimental ones reported in the literature. Note that since the results vary randomly from one realization to another one, one cannot expect agreement in detail between a single run of the simulation and a single experimental discharge.

10.3 Prediction of the 50% negative breakdown voltage

The input data are the geometry of electrodes, the wave shape of the voltage, and the ambient air pressure and temperature, given that this model is a self-consistent time-dependent one. Comparisons are made with the test results reported by the Les Renardières Group [1]. The experimental results concern U_{50} voltage determination for rod–plane gap distances varying from 2.6–10 m. The rod consisted of a hemispherical electrode tip 0.6 m in diameter. A non-standard negative switching surge of 60/3000 μs was considered. The U_{50} voltage was computed for each gap distance from the model (2.6, 3, 4, 7, and 10 m), considering a total of 50 shots. The up-and-down testing procedure was used.

Table 10.1 summarizes the simulated values of U_{50} for 50 shots and for each gap. It appears that good agreement exists between the simulated values of U_{50} and the experimental ones obtained under the same experimental conditions as indicated in table 10.1.

The simulated and experimental results were plotted in the same graph for better comparison (figure 10.3).

The proposed model is effective because the mean of the standard deviation of the modeled output of the U_{50} is below 5%. The simulated dispersions range from 2.4%–4.2%. These values also fit well with the experimental ones, where the

Table 10.1. Comparison of simulated and experimental values of U_{50}. σ represents the standard deviation of the simulated results.

Electrode gap D (m)	2.6	3	4	7	10
Experimental values of U_{50} (kV) [1]	1890.5	1994.8	2302	3017.3	3464.3
Simulated values of U_{50} (kV)	1813.9	1934.6	2202	2832.5	3325.7
σ (%)	2.4	3.0	3.4	3.9	4.2

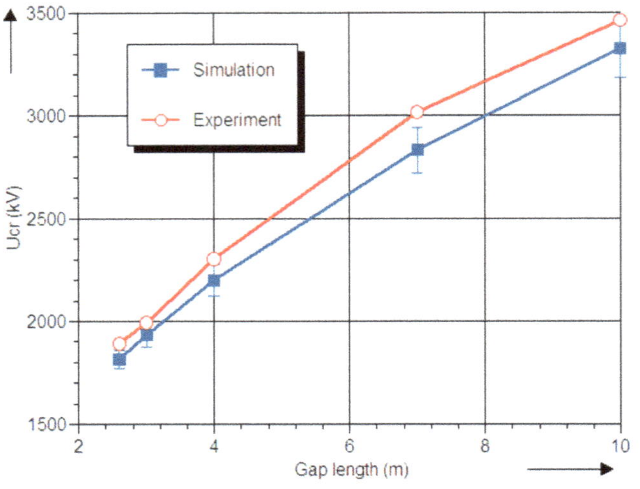

Figure 10.3. Comparative summary of the U_{50} simulated results with experimental data reported by the Les Renardières Group [1]. The U_{50} voltage was determined as a function of the gap length D (electrode radius R_p = 0.6 m, –60/3000 μs negative wave).

dispersions vary between 4%–5% [1]. These results are in satisfactory agreement, although the margin of error tends to increase for larger gap lengths. The maximum error (for $D = 7$ m) is about 6.1%.

Negative discharge phenomena are so complex that many simplifying assumptions have to be made in order to render mathematical modelling possible. The error in the U_{50} prediction may therefore be related to simplifying assumptions, the influence of humidity, and/or the empirical relationships derived from laboratory experiments.

Note that since the influence of humidity is not actually considered in the models, U_{50} may be determined under normal humidity conditions with correction factors being applied. Weather correction factors can be applied according to the IEEE Standard [8] instead of the more recent IEC Standard [9]. The IEEE Standard is accurate over a wider range of gap geometries and spacings than the IEC Standard. For extremely long air gaps the IEC correction is much smaller than the appropriate correction [10].

References

[1] Les Renardières Groupe 1981 *Electra* **74** 67

[2] Rakotonandrasana J H *et al* 2008 *J. Phys. D: Appl. Phys.* **41** 105210

[3] Ortéga P 1992 Comportement diélectrique des grands intervalles d'air soumis à des ondes de tension de polarité positive ou négative *PhD Thesis* Université de Pau et des Pays de l'Adour France

[4] Rees T 1997 Contribution à l'étude expérimentale et théorique des décharge électriques de polarité négative dans l'air à pression atmosphérique *PhD Thesis* Université de Pau et des Pays de l'Adour, France

[5] Bazelyan E M, Raizer Yu P and Aleksandrov N L 2007 *J. Phys. D: Appl. Phys.* **40** 4133

[6] Rizk F A M 1989 *IEEE Trans. Power Deliv* **4** 596
[7] Hutzler B and Hutzler D 1992 *Bull. Dir. Etud. Rech* **4** 11
[8] IEEE 1995 *IEEE Standard Techniques for High Voltage Testing* IEEE Std 4-1995
[9] IEC 2006 *High-voltage Test Techniques. Part 3: Definitions and Requirements for On-site Testing* (London: IEC) 60060-3
[10] Hill K H 1999 Estimation of positive polarity switching surge flashover voltage of compact transmission lines *MSc Thesis* University of Manitoba, Canada

Chapter 11

Application of the model to positive lightning discharge

11.1 Introduction

Lightning is a transient high current electric discharge that occurs over extreme distances when some regions of the atmosphere, generally part of the thundercloud, attain an electric charge sufficiently larger than the air's dielectric strength. Two kinds of dangerous effects upon engineering systems should be emphasized. The first one is the direct hit: the discharge occurs when the discharge strikes the device directly. This case is the most dangerous, but it is fortunately not the most common. The second effect is due to the radiated electric fields from both the leader and return strokes. Indeed, the accompanying bolt of current radiates strong electric and magnetic fields that can also have devastating effects on engineering systems (communications devices, electronics integrated into different systems, navigational computers, and so forth).

Most cloud-to-ground lightning in temperate regions brings down negative charges to Earth. Flashes that drop positive charges to the ground are generally referred as positive lightning. In spite of its relatively sporadic nature, making up only 10% to about 30% of strikes in temperate zones, positive lightning is of particular interest for two reasons: (i) it is responsible for the largest recorded lightning currents, those in the 200–300 kA range [1–3], and for charges transferred to Earth that are considered larger than those of negative flashes; and (ii) in some tropical zones, even if statistics are lacking, the occurrence of positive lightning seems to be about 80% of total lightning strikes [1].

In this chapter simulation results obtained from the mathematical model described in chapter 7 are provided; simulation results showing the influence of soil conductivity and cloud-to-ground distance on the leader and return stroke currents are also presented. In addition to lightning parameters (discharge current, propagation velocity, etc), radiated electric field changes of the leader and return

doi:10.1088/978-0-7503-1236-3ch11

strokes as well as the magnetic field changes associated with the leader are also evaluated. The computed electromagnetic fields radiated by the leader and the return stroke are used in this chapter to study their effects on engineering systems. It is also shown that, contrary to studies performed using lightning leader electromagnetic field calculations [1, 4], the leader linear distributions of charge and velocity are not constant but time-dependent.

11.2 Prediction of positive lightning discharge parameters

The mathematical model presented in chapter 7 to describe stepped positive lightning was used to predict the lightning discharge parameters. The cloud base is simulated by a large flat surface at a given potential located at a height D (m) above the ground. This arrangement provides a uniform electric field between the cloud and the ground.

Consider for example a cloud–ground distance $D = 6000$ m; this value is in the range of those used in the literature [1–4]. Assuming that the field distribution is roughly uniform between the cloud base and the Earth, the voltage drop between them can be estimated (before the beginning of the preliminary discharge) to be a few tens of megavolts. Therefore, a 6000 m gap length subjected roughly to a constant voltage of about $U_c = 65$ MV under specific atmospheric conditions of $P = 100$ kPa and $T = 21$ °C at ground level. For the cloud diameter, a kilometre constitutes a suitable dimension value [1–3].

Figures 11.1–11.6 show the simulated results of the lightning's downward leader characteristics. The leader plausible trajectory is shown in figure 11.1. The erratic

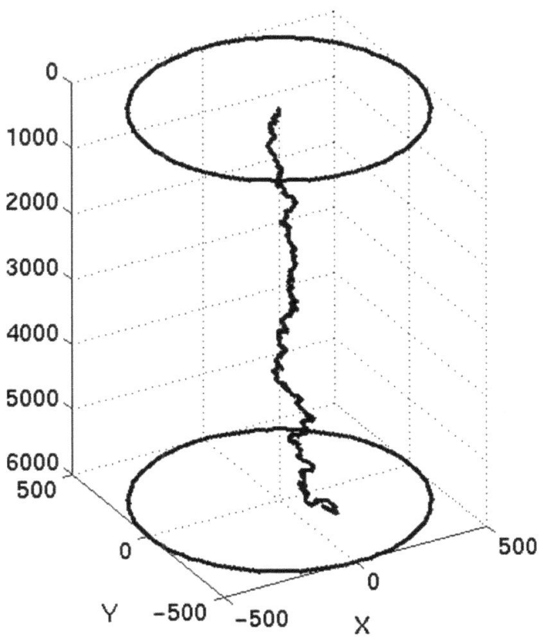

Figure 11.1. Simulation of the leader trajectory in the space reprinted from [5] with permission from Elsevier.

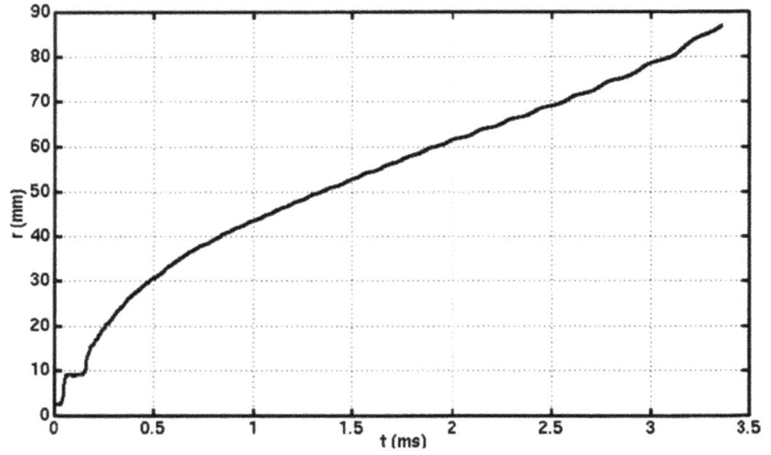

Figure 11.2. The leader's thermal radius evolution reprinted from [5] with permission from Elsevier.

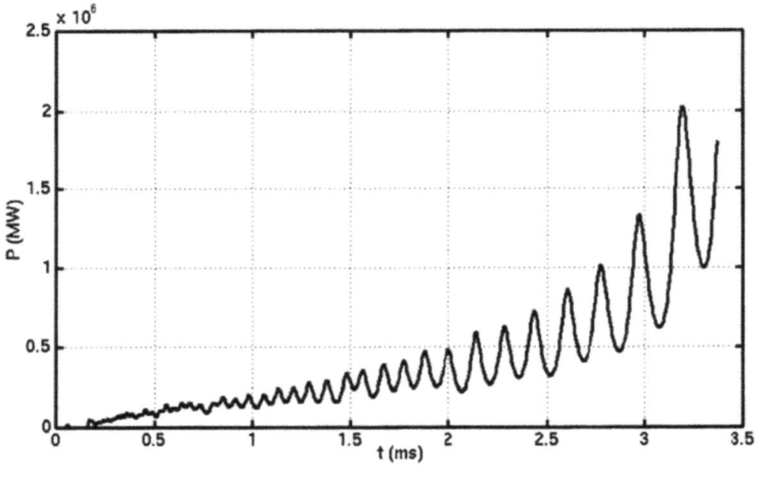

Figure 11.3. Power injected in the gap reprinted from [5] with permission from Elsevier.

propagation of the lightning discharge path can be recognized. The characteristics of its thermal radius are summarized in figure 11.2. Its growth reaches a few centimetres, as observed in natural lightning [3, 7, 8]. The power and energy dissipated during the leader propagation are shown in figures 11.3 and 11.4, respectively. Figure 11.5 depicts the leader current. Its mean value is about 2–10 kA while recorded leader currents averaged in the range of 0.2–10 kA [1–3]. The corresponding charge drained to Earth is presented in figure 11.6. The peak value reaches a few tens of coulombs; this value is in the order of the experimental results [1].

Figure 11.7 depicts the leader's instantaneous velocity where the average velocity is about 2×10^6 m s^{-1} which is in good agreement with the results recorded by

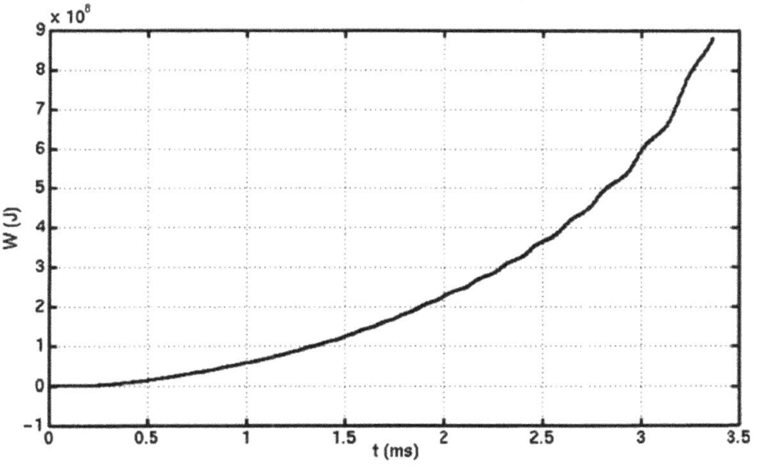

Figure 11.4. Energy injected in the gap reprinted from [5] with permission from Elsevier.

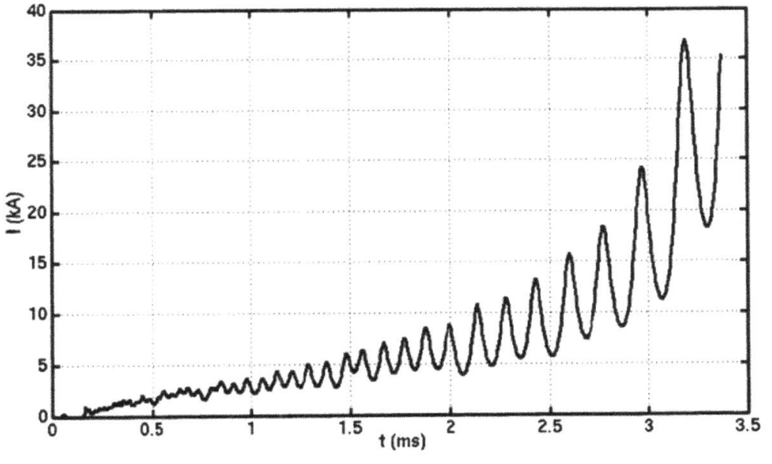

Figure 11.5. Leader current reprinted from [5] with permission from Elsevier.

Berger [9] for positive lightning. Indeed, Berger [9] obtained 3.6×10^5 m s^{-1} near the cloud base and 2.4×10^6 m s^{-1} near the Earth as the leader velocity and an average value of about 2×10^6 m s^{-1}. Note also that Rubinstein *et al* [10] have obtained leader speeds between 2×10^6 m s^{-1} to 2×10^7 m s^{-1} for negative flashes.

11.3 Influence of soil conductivity and cloud–ground distance on the positive lightning impulse current

A lightning flash's density depends to some extent on the electro-geological properties of the soil and its topography (terrain type, slope, drainage, soil saturation reflected by conductivity, etc) as well as the structure on the ground [11]. The knowledge of the

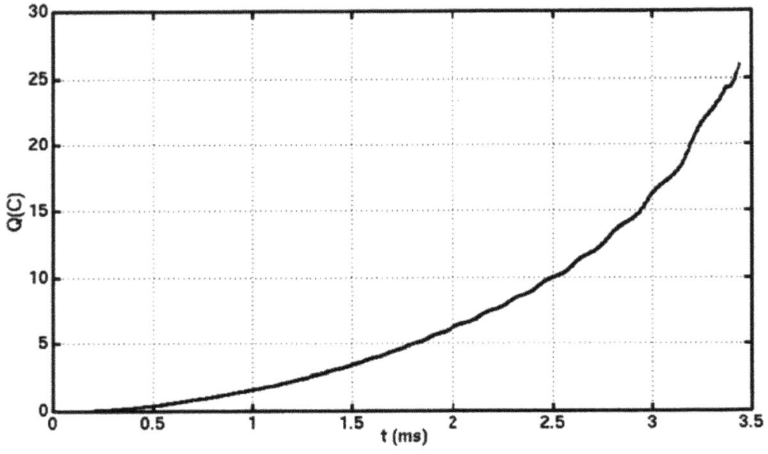

Figure 11.6. Leader charge injected in the gap reprinted from [5] with permission from Elsevier.

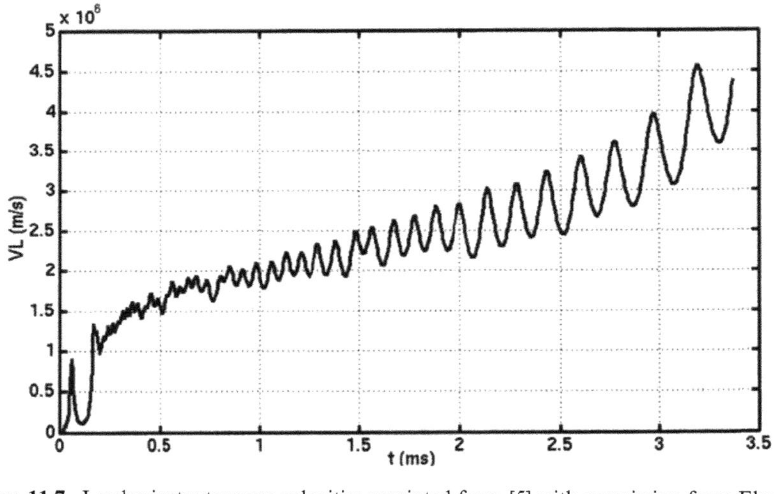

Figure 11.7. Leader instantaneous velocities reprinted from [5] with permission from Elsevier.

density of lightning flashes is very important for the design and dimensioning of lightning protection. Thus, the need for lightning prediction is clear; many real-time lightning detection systems are now able to accurately determine the impact location of cloud–ground lightning.

Many experimental studies have been carried out on the effectiveness of lightning protection systems (i.e. the capture zone) as a function of soil conductivity. It was reported that protection effectiveness increases with decreasing soil conductivity [12–16].

In this section, the influence of soil conductivity and cloud–ground distance on the positive lightning discharge characteristics is emphasized. The soil is assumed to be homogeneous, isotropic, and non-ferromagnetic (i.e. the relative permeability is taken as equal to one) with a uniform conductivity. A wide range of conductivities reflecting different types of soils were considered: 1–200 μS cm^{-1}. In reality, these assumptions are far from the truth because of the composition, chemistry, and moisture content of different soil geologies.

The cloud–ground distances considered are 2000, 3000, 4000, 6000, and 7500 m. The obtained results are discussed in the light of the experimental data reported in the literature.

11.3.1 The influence of soil conductivity

Consider a square area of ground 1000 m^2 wide and 1 m thick, together with a cloud diameter of 1000 m. Figures 11.8–11.10 depict the leader current, the injected electrical charge, and the return stroke current, respectively, for a cloud–ground distance D = 2000 m, subjected to a constant voltage U = 35 MV and for three different values of soil conductivity (σ). It may be observed that any increase in the soil conductivity induces an increase in the leader and return stroke currents with a concomitant decrease in the time to breakdown t_b. Apparently, a large soil conductivity might increase lightning flash density. These findings have been reported by other investigators [12].

Thus, according to observations by many investigators, lightning discharges have a preferential attraction to swamp ground, and humid soil and clothes [3]. The presence of water increases soil conductivity. It can, therefore, be expected that any conducting structures located on highly resistive ground might be attractive targets for lightning discharges.

The average velocity (V_{aver}) of the lightning discharge is also influenced by the soil conductivity (σ); V_{aver} increases with σ (table 11.1).

Consider another example, of a cloud–ground distance D = 3000 m and soil conductivities taken as 100 and 200 μS cm^{-1}. Figures 11.11 and 11.12 show, the variation of the leader current, the injected electrical charge, and the return stroke current of lightning discharges. Again, it may be observed that the currents and electrical charge increase with soil conductivity, while the time to breakdown decreases when σ increases.

11.3.2 Influence of cloud–ground distance

The influence of cloud–ground distance was investigated considering the same soil conductivity σ = 100 μS cm^{-1}. Some of the main lightning characteristics are reported in figures 11.13 and 11.14. It may be observed that the average velocity, the height of final jump, and the time to breakdown increase with D (table 11.2). It should be emphasized that the characteristics predicted by the model are in good agreement with those recorded during natural lightning flashes [1]. This agreement includes the shape as well as the amplitude of the parameters characterizing the discharges.

11.4 Electric field changes of the leader and return stroke

A cloud–ground distance D = 6000 m was considered, with a constant voltage U_c = 70 MV, under specific atmospheric conditions of P = 100 kPa and T = 25 °C,

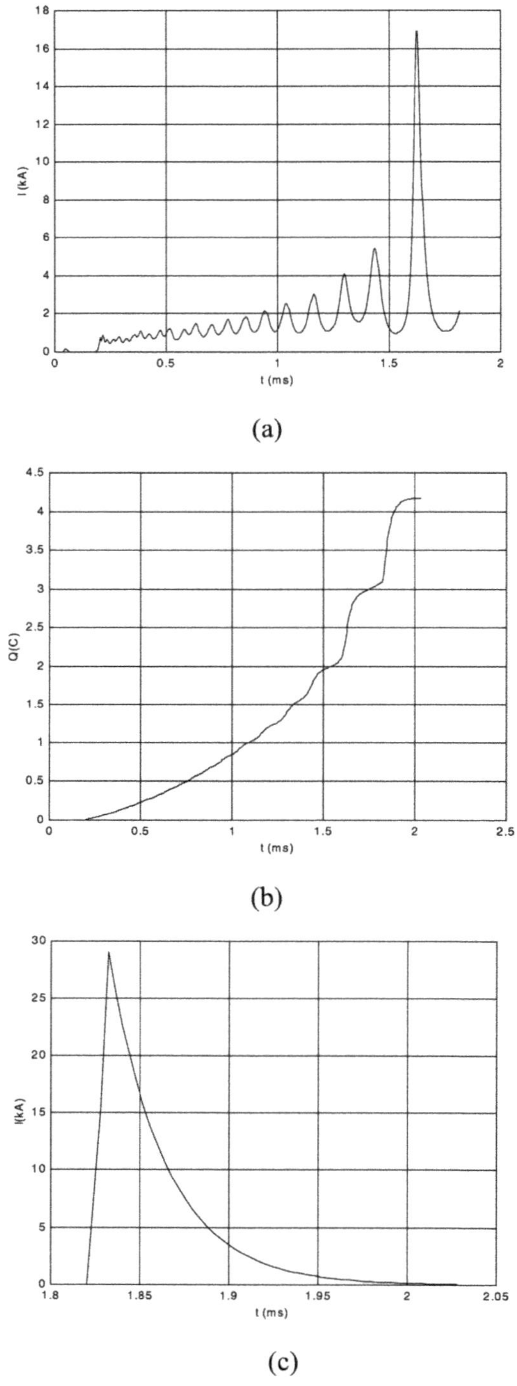

Figure 11.8. Lightning discharge parameters for $D = 2000$ m and soil conductivity $\sigma = 100\ \mu\text{S cm}^{-1}$. (a) Leader current, (b) charge injected in the gap, and (c) return stroke current.

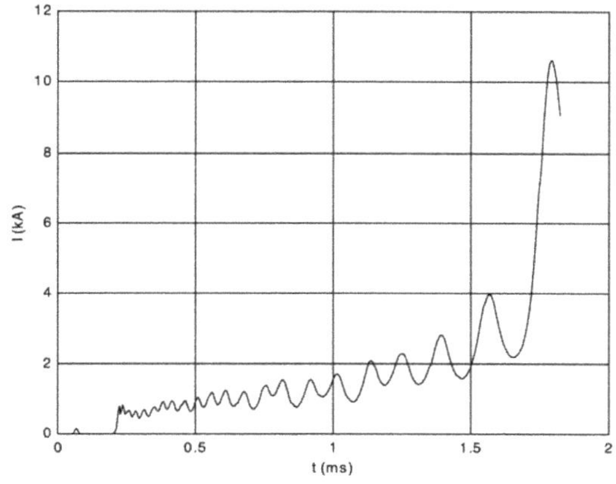

(a)

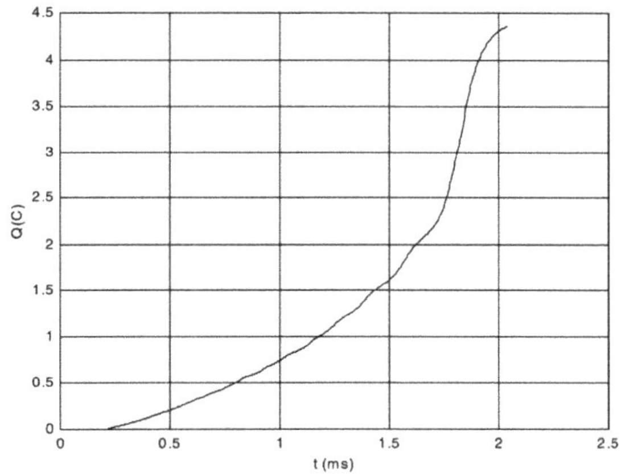

(b)

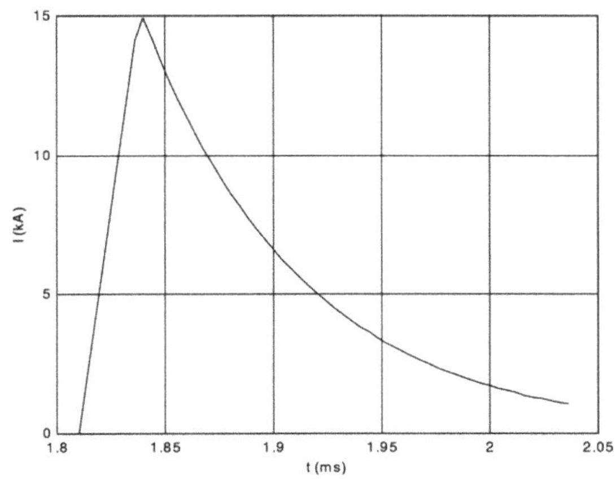

(c)

Figure 11.9. Lightning discharge parameters for $D = 2000$ m and soil conductivity $\sigma = 20\ \mu\text{S cm}^{-1}$. (a) Leader current, (b) charge injected in the gap, and (c) return stroke current.

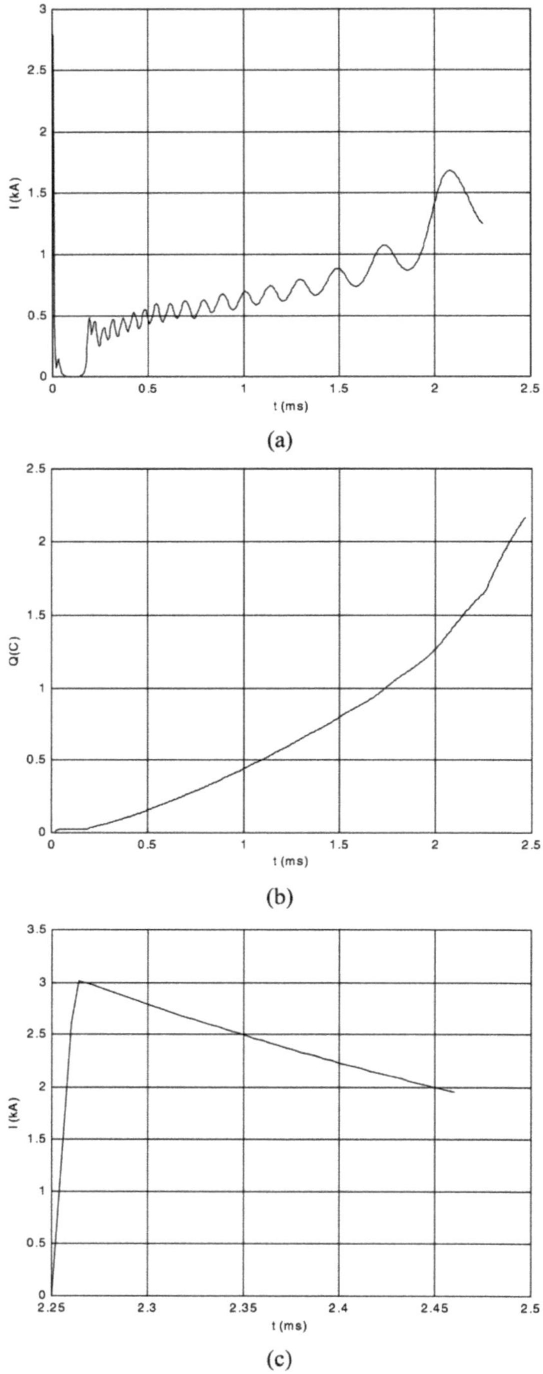

Figure 11.10. Lightning discharge parameters for $D = 2000$ m and soil conductivity $\sigma = 2\ \mu\mathrm{S\ cm}^{-1}$. (a) Leader current, (b) charge injected in the gap, and (c) return stroke current.

Table 11.1. The influence of the soil conductivity on the average velocity (V_{aver}) of lightning discharge, the height of the final jump (H), and the time to breakdown t_b.

σ (μS cm^{-1})	V_{aver} (10^6 m s^{-1})	H (m)	t_b (ms)
1	0.7808	79.2	2.5
2	0.8621	79.2	2.3
5	0.958	79.2	2.1
10	1.0174	79.2	1.9
20	1.0743	79.2	1.8
50	1.0743	79.2	1.8
100	1.0695	79.2	1.8

at the ground level. A cloud diameter of a kilometre is used; this value constitutes a good dimension order [1, 5, 17–24].

The vertical and horizontal field changes at 50–550 m from the strike point are represented in figures 11.15 and 11.16. Looking at these waveforms, two main phases can easily be distinguished. The first one recorded during the first 50 μs corresponds to the leader electric field change. The electric field at ground level increases with leader advancement until the beginning of the return stroke. This increase corresponds to the peak in the electric fields. During the return stroke phase, the electric field decays due to the charge neutralization in the channel.

In the literature, no experimental data are given for the positive ground flashes having such characteristics. However, according to Uman [1], electromagnetic fields from positive flashes are apparently similar to fields from negatives flashes except for their polarity [26]. To provide a base line for comparison, a waveform of the vertical electric field measured in 1993 during a triggered lightning campaign [27] is given in figure 11.17. This 50 m field waveform exhibits an asymmetrical V-shape as obtained by the proposed model. It can be noticed that the calculated values represent a good approximation of the measured values, except for the polarity.

11.5 Magnetic field associated with the leader

In addition to the electrostatic aspect of the leader, its current, which is a few kA [1–4, 28], is enhanced at ground level due to a variation of the magnetic field. Figure 11.18 shows a simplified geometric model allowing for the calculation of this field at M (x_0, y_0, z_0).

The same hypotheses as addressed in the previous section were considered. The channel is subdivided into successive current elements $I\mathrm{d}z$. The contribution of each element is [25, 29]:

$$\mathrm{d}H = \frac{I\mathrm{d}z}{4\pi}\frac{Dh}{(z^2 + Dh^2)^{\frac{3}{2}}}. \tag{11.1}$$

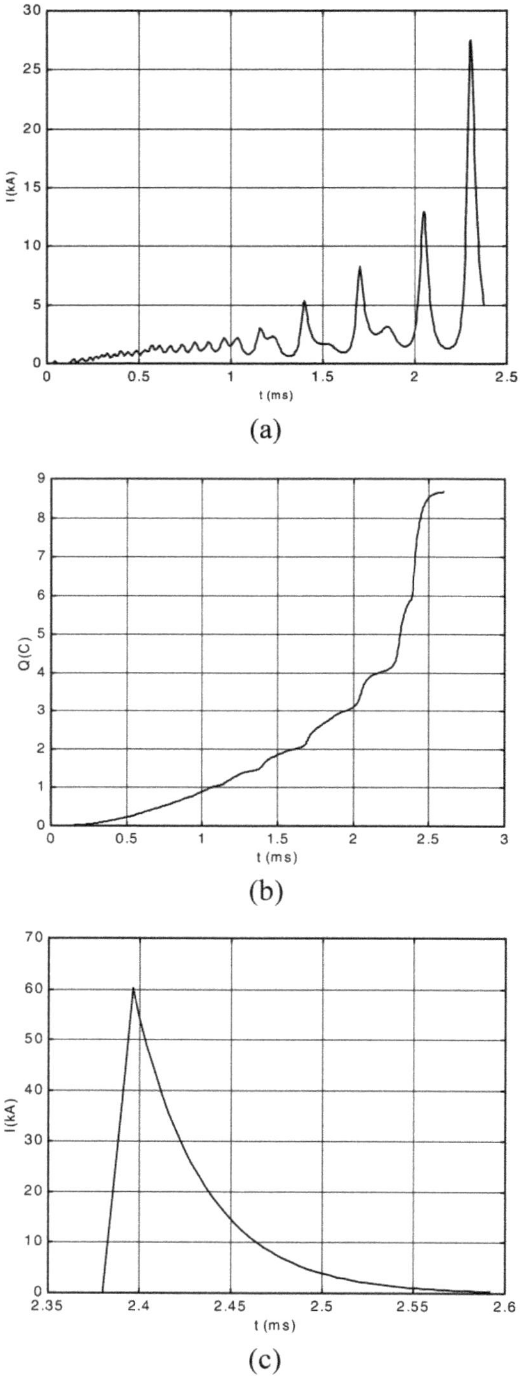

Figure 11.11. Lightning discharge parameters for D = 3000 m and soil conductivity σ = 100 $\mu S\ cm^{-1}$. (a) Leader current, (b) charge injected in the gap, and (c) return stroke current.

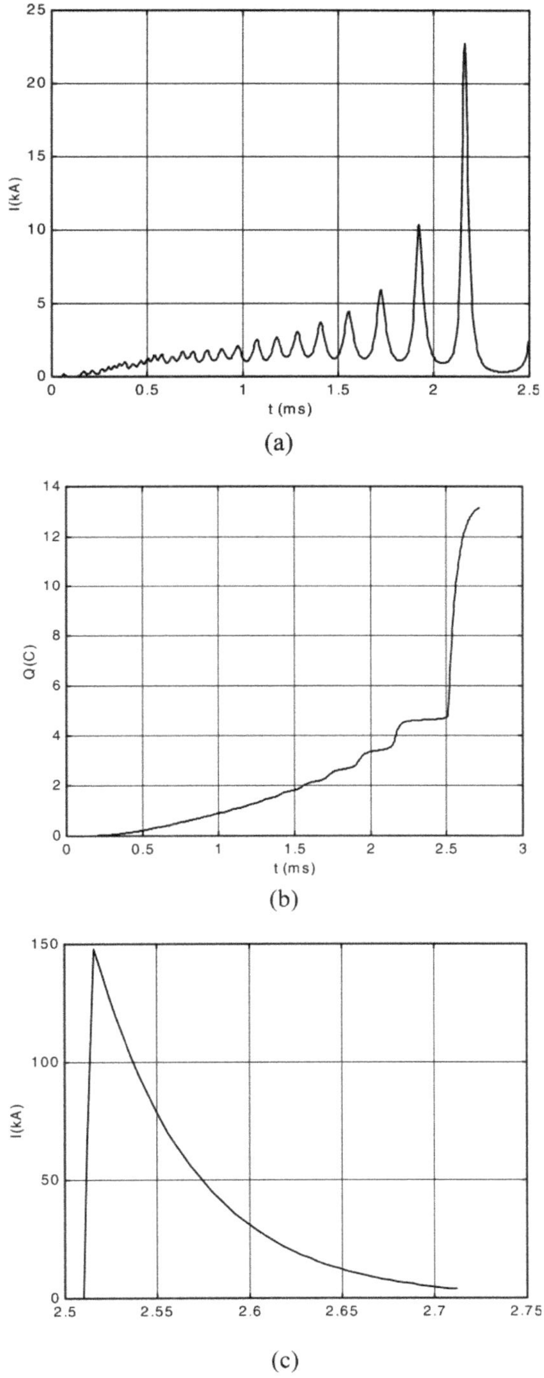

Figure 11.12. Lightning discharge parameters for D = 2000 m and soil conductivity σ = 200 μS cm^{-1}. (a) Leader current, (b) charge injected in the gap, and (c) return stroke current.

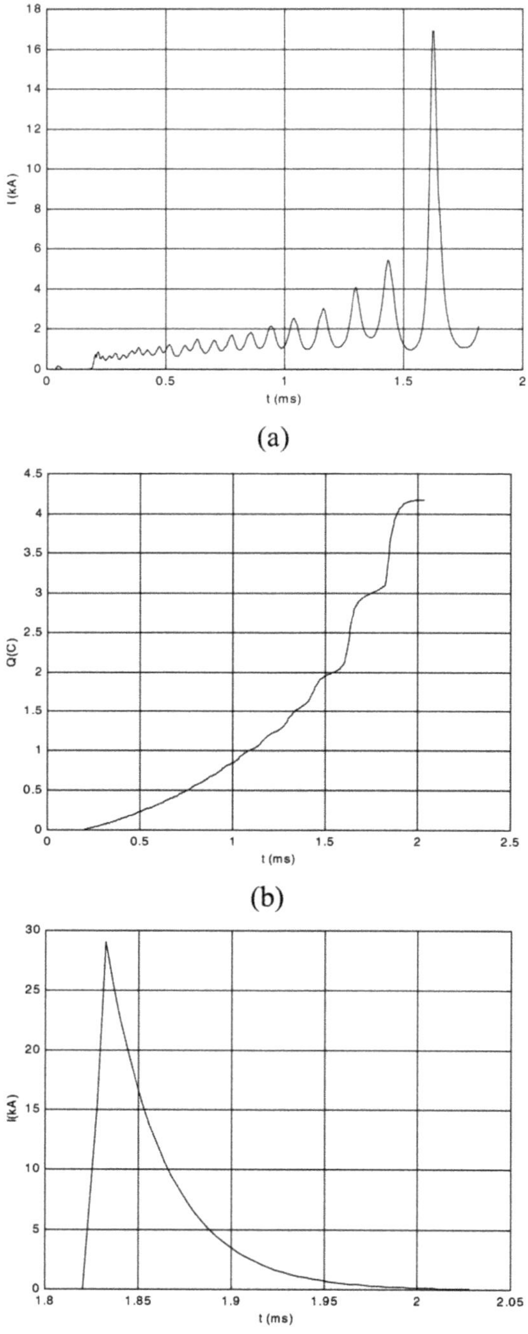

Figure 11.13. Lightning discharge parameters for $D = 2000$ m. (a) Leader current, (b) charge injected in the gap, and (c) return stroke current.

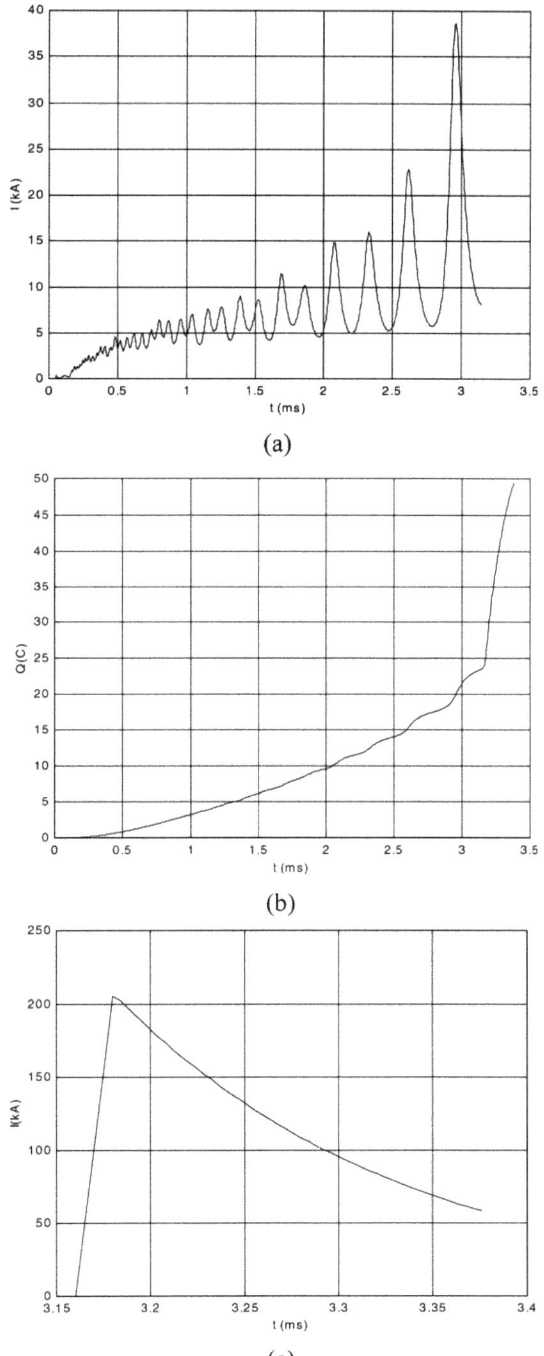

Figure 11.14. Lightning discharge parameters for D = 7500 m. (a) Leader current, (b) charge injected in the gap, and (c) return stroke current.

Table 11.2. Influence of the cloud–ground distance D on the average velocity (V_{aver}) of lightning discharge, the height of the final jump (H), and the time to breakdown t_b. U is the applied voltage.

D (m)	U (MV)	V_{aver} (m s^{-1})	H (m)	t_b (ms)
2000	35	1.0965×10^6	79.2	1.8
4000	45	1.4×10^6	101.93	2.8
6000	60	1.95×10^6	136.02	3.1
7500	75	2.35×10^6	158.75	3.2

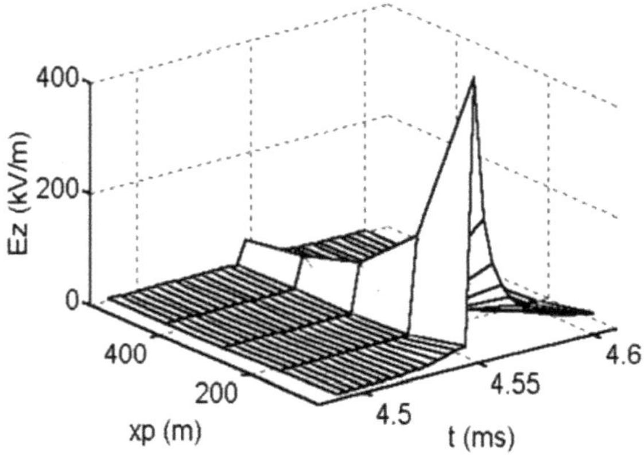

Figure 11.15. The vertical electric field distribution of the discharge at 50–550 m from the strike point.

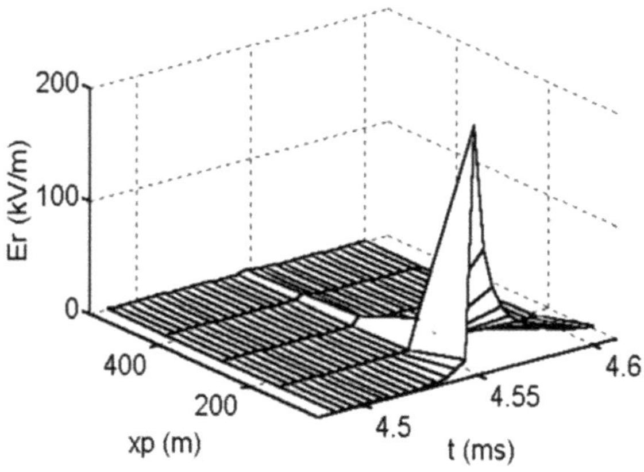

Figure 11.16. The horizontal electric field distribution of the discharge at 50–550 m from the strike point.

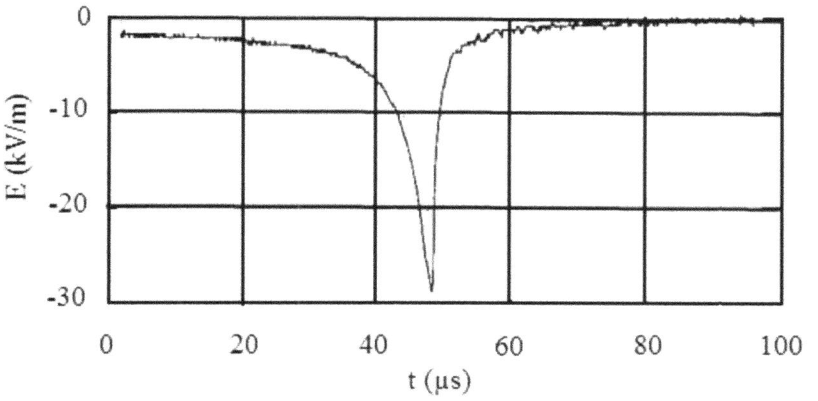

Figure 11.17. Measured vertical electric field waveforms at 50 m, corresponding to the 5th return stroke. Reproduced from [4] with permission of Ecole Central de Lyon, France.

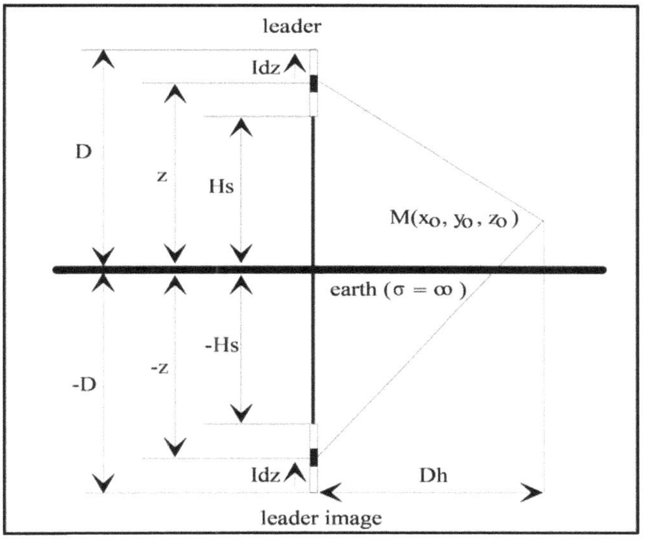

Figure 11.18. Simplified geometric model used for the leader magnetic field calculation [4, 5]. Reproduced from [4] with permission of Ecole Central de Lyon, France.

The field due to the total leader length at the time t is obtained by integrating the above relation on this length. The total field due to the leader and its image contribution, at the point M is as follows:

$$H = \frac{I}{4\pi Dh}\left[\begin{array}{l} \dfrac{D - z_o}{\left((D - z_o)^2 + Dh^2\right)^{\frac{1}{2}}} - \dfrac{Hs - z_o}{\left((Hs - z_o)^2 + Dh^2\right)^{\frac{1}{2}}} \\ + \dfrac{D + z_o}{\left((D + z_o)^2 + Dh^2\right)^{\frac{1}{2}}} - \dfrac{Hs + z_o}{\left((Hs + z_o)^2 + Dh^2\right)^{\frac{1}{2}}} \end{array}\right] \tag{11.2}$$

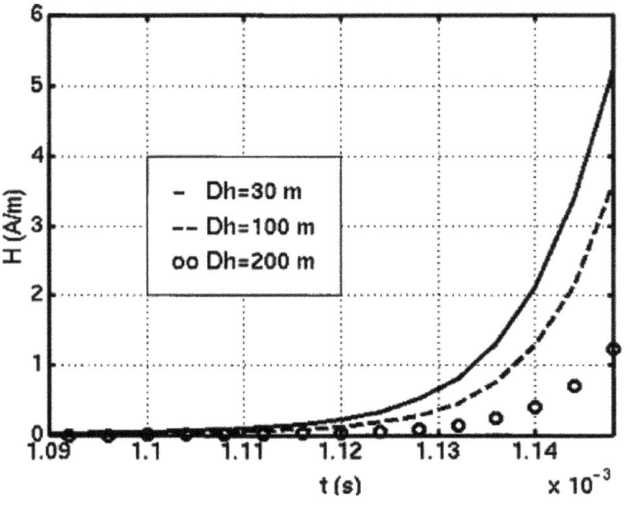

(a)

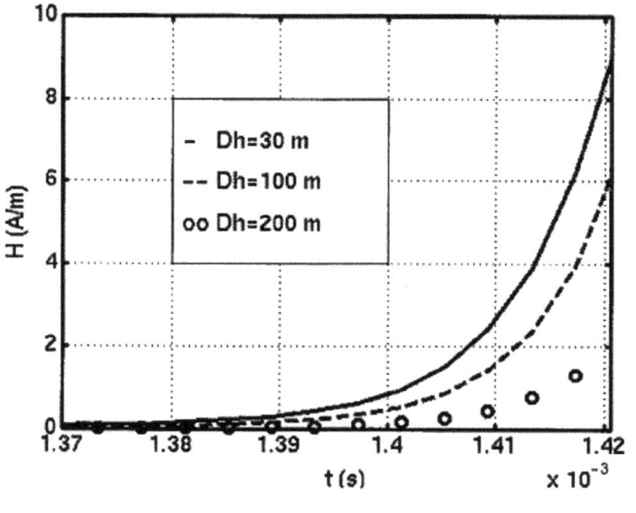

(b)

Figure 11.19. Magnetic field changes of the discharge 50 μs before the return stroke, at three different positions from the discharge axis. (a) $D = 5000$ m and (b) $D = 7500$ m.

Simulations were performed for two different cloud–ground distances, with D equal to 5000 m and 7500 m, respectively. The magnetic field changes recorded 50 μs before the return stroke at three different positions from the discharge axis are illustrated in figure 11.19.

The amplitude and waveform of the predicted values are similar to those measured in 1993 at Camp Blanding, Florida during a triggered lightning campaign [30].

References

[1] Uman M A 1987 *The Lightning Discharge (International Geophysics Series vol 39)* (New York: Academic)
[2] Aguet M and Ianoz M 1987 *Haute Tension* (Paris: Dunod)
[3] Le Roy G, Gary C, Hutzler B, Lalot J and Dubanton C 1984 *Les Propriétés Diélectriques de l'Air et les Très Hautes Tensions (Collection de la Direction des Études et Recherches d'Électricité de France)* vol 51 (Paris: Eyrolles)
[4] Ben Rhouma A 1996 Caractérisation et modélisation des champs électromagnétiques rayonnés par les décharges de foudre en vue de l'élaboration d'un simulateur expérimental *PhD Thesis* Ecole Centrale de Lyon, University of Lyon
[5] Fofana I and Beroual A 1998 *Electr. Power Syst. Res.* **48** 105
[6] Noskov M D, Pleshkov O I and Lopatin V V 2000 Computer modelling of earth conductivity and relief effects on lightning orientation *25th Int. Conf. Lightning Protection (Rhodes, Greece,* 18–22 *September)* p 389
[7] Zeddam A and Degauque P 1987 *Electromagnetics* **7** 541
[8] Degauque P and Zeddam A 1988 *IEEE Trans. Electromagn. Compat.* **30** 77
[9] Beger K 1967 *J. Franklin Inst.* **283** 478
[10] Rubinstein M, Uman M A, Medelius P J and Thomson E M 1994 *IEEE Trans. EMC* **36** 134
[11] Anderson R B and Erikson A J 1990 *Electra* **69** 65
[12] Nixon K J, Jandrell I R and Geldenhuys H J 2000 Sensitivity analysis of parameters influencing the lightning impulse current behaviour of earth electrode components *25th Int. Conf. Lightning Protection (Rhodes, Greece,* 18–22 *September*) 484
[13] Szpor S 1971 *Arch. Elektrotech.* **20** 379
[14] Szpor S 1973 *Arch. Elektrotechn.* **22** 735
[15] Szpor S 1974 *Arch. Elektrotchn.* **23** 1001
[16] Noskov M D, Pleshkov O I and Lopatin V V 2000 Computer modelling of earth conductivity and relief effects on lightning orientation *25th Int. Conf. Lightning Protection (Rhodes, Greece,* 18–22 *September*) p 389
[17] Diendorfer G 1990 *IEEE Trans. EMC* **32** 292
[18] Rachidi F, Rubinstein M, Nucci C A and Guerrieri S 1994 Influence of the leader electric field change on voltages induced by very close lightning on overhead lines *22nd Int. Conf. Lightning Protection (Budapest, Hungary,* 19–23 *September)* paper R 3b-10
[19] Georgiadis N, Rubinstein M, Uman M A, Medelius P J and Thomson E M 1992 *IEEE Trans. Electromagn. Compat.* **34** 451
[20] Yokohama S 1984 *IEEE Trans. Power Appar. Syst.* **103** 100
[21] Cooray V and De La Rosa F 1986 *IEEE Trans. Antennas Propag.* **34** 88
[22] Zeddam A and Degauque P 1990 Current and voltage induced on telecommunication cable by a lightning stroke *Lightning Electromagnetics* ed R L Gardner (Washington, DC: Hemisphere) 377
[23] Rubinstein M, Tzeng A Y, Uman M A, Medelius P J and Thomson E M 1989 *IEEE Trans. EMC* **31** 376
[24] Rachidi F, Nucci C A, Ianoz M V and Mazzetti C 1996 *IEEE Trans. EMC* **38** 250
[25] Fofana I, Beroual A, Rhouma B A and Auriol P 1998 *IEEE Power Gen. Trans. Syst.* **145** 395

[26] Rakov V A, Uman M A, Fernandez M I, Thottappillil R, Eybert-Berard A, Berlandis J P, Rachidi F, Rubinstein M, Guerrieri S and Nucci C A 1996 Observed electromagnetic environment close to the lightning channel *Proc. 23rd Int. Conf. on Lightning Protection (Firenze, Italy, 23–27 September)* p 30

[27] Master M J, Uman M A, Beasley W H and Darveniza M 1986 *IEEE Trans. EMC* **28** 158

[28] Schonland F J 1956 *The Lightning Discharge (Handbuch der Physik vol 22)* (Berlin: Springer)

[29] Cheng D K 1983 *Field and Wave Electromagnetics* (Reading, MA: Addison-Wesley)

[30] Uman M A, Rakov V A, Rambo K J, Vaught T W, Fernandez M I, Cordier D J, Chandler R M, Bernstein R and Golden C 1997 *Trans. IEEE Japan* **117B** 446

Chapter 12

Application of the model to the process of lightning–ground connection and quantification of the striking distance

12.1 Introduction

This chapter presents an analytical approach to model the attachment process of a negative downward stepped leader to a ground structure. The impact, on a tall ground structure, is determined by the connection between a downward stepped leader and an upward discharge growing up from this structure. This upward discharge is a part of the transition from the leader to the return stroke stage [1]. Since the attachment process is important for the understanding of lightning physics and it is also fundamental for lightning protection techniques, it has received considerable attention in the last few decades [2–4].

When the field strength between the ground and the cloud becomes large enough, a pilot leader is initiated. This downward stepped leader rushes towards the ground ionizing a path in the air. As the stepped leader approaches the ground, the electric fields at the extremities of tall grounded objects increase to such a level that they may launch connecting leaders towards the descending stepped leader. The first return stroke is initiated at the instant of connection between the descending stepped leader and one of the connecting leaders. The successful launching of a connecting leader requires the inception of a streamer discharge from the tall grounded object (sharp protrusion, rod, tower, tree, etc) and its transition into a leader (i.e. streamer–leader transition). The upward connecting discharge is a part of the transition from the leader stage to the return stroke stage, this transition being termed the lightning attachment process.

This approach can be helpful in the predetermination of lightning protection parameters such as the impact points and the striking distance of lightning. The results of the sensitivity analysis of an upward leader initiation from a rod, in response to an approaching lightning downward leader, is explored in this chapter.

12.2 Modelling the lightning connection process to a ground structure

In this section, the progression of an upward discharge, in the direction of the negative leader, is assumed for an air gap breakdown between the structure and the tip of the downward leader. The model outlined in chapter 7 is considered hereafter; the negative downward leader being regarded as a uniformly charged line propagating axially.

Taking into account some atmospheric conditions, the model allows the determination of some parameters of the upward leader, such as the current, the corresponding charge, the propagation velocity, the trajectory, and the power and energy injected into the gap. The computed results are compared with the experimental data obtained during a triggered lightning campaign performed at Camp Blanding, in the summer of 1995 [3, 4].

12.2.1 Background on lightning connection process

Recall from chapter 11 that when the electric field strength between the ground and a thunderstorm cloud becomes large enough, a pilot leader is initiated. This downward negative stepped leader rushes toward the ground ionizing a path in the air. When its tip approaches some tens of metres to the ground, an upward positive connecting discharge is initiated from the ground or a tall ground structure to connect to the downward leader. The upward connecting discharge is a part of the transition from the leader stage to the return stroke stage; this transition is termed the lightning attachment process. Investigations at Camp Blanding [1] reported that the return stroke is inferred to start at the point of junction elevated above the ground, between the downward dart leader and the upward connecting discharge and propagating in both upward and downward directions. Such a finding was previously hypothesized by a number of researchers [5, 6]. A schematic time-resolved picture of the attachment process is given in figure 12.1.

12.2.2 General description of the modelling

The model of the discharge processes depicted by the general flow chart of figure 12.2 is a self-consistent time-dependent model. The simulation code is composed of three major coupled sections.

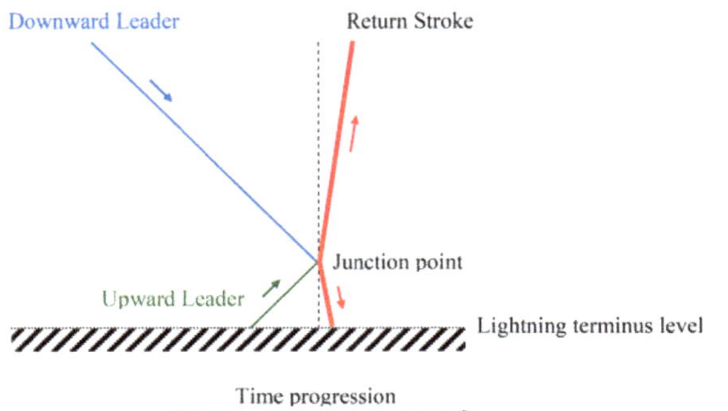

Figure 12.1. A streaked image sketch of the attachment process of an artificially triggered lightning sequence [1].

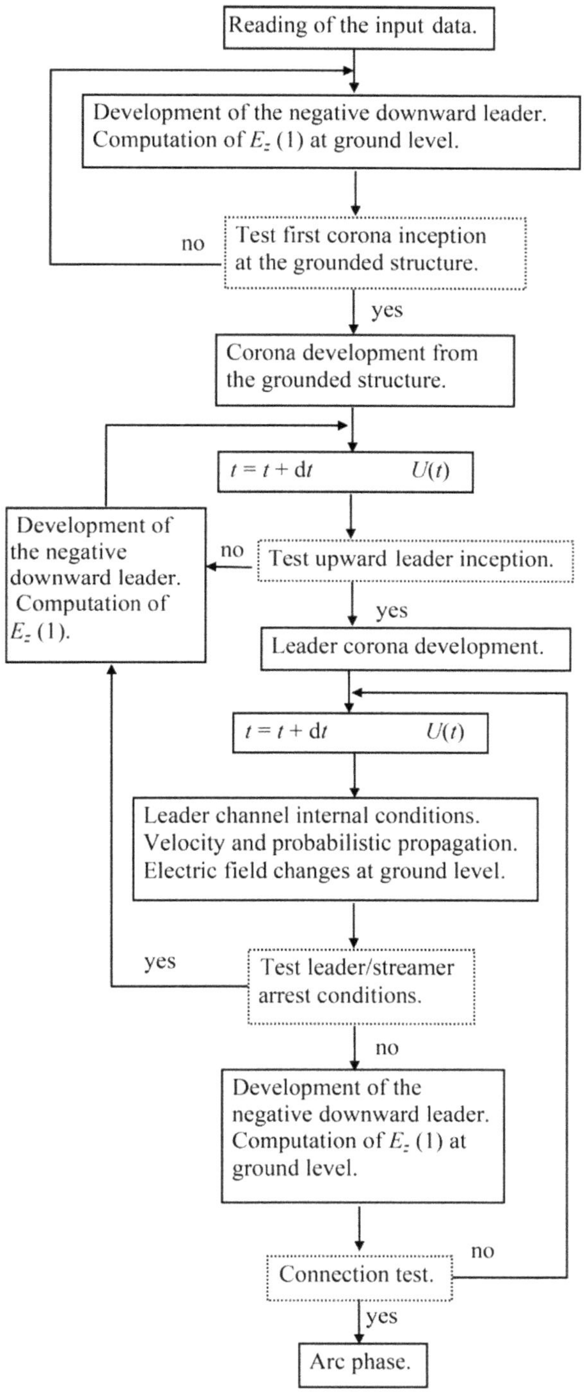

Figure 12.2. General flow-chart of the modelling.

The negative downward stepped leader

The negative downward leader is modelled by considering a simplified electrostatic model. The channel is regarded as a uniformly charged line (75 μC m^{-1}) propagating axially towards the ground or a ground structure at a constant velocity of about 4.5×10^5 ms^{-1} [3]. The negative leader initially propagates without the influence of the positive leader. Its progression enhances the electric field at ground level.

Just before the final jump of the downward negative leader, one or several upward positive discharges are initiated, propagating towards the negative leader, given a certain critical field is reached. These upward positive discharges are constituted by leaders at the tip of which positive streamers are generated [7]. The progression of the upward discharge is, therefore, assumed to be an air gap breakdown between the tall grounded structure and the tip of the downward negative leader.

The positive upward leader

The mathematical model considered for the upward discharge is the one presented in chapter 7. This model is used based on the substantial similarity existing between lightning and discharges in long air gaps. Recall that in this model, the discharge channel is represented by an equivalent electrical network, the parameters of which are time-dependent and determined by physical laws, gas discharge theory, and the discharge geometry. The input data are the electrode geometry, voltage waveform, and ambient air pressure and temperature. In the calculations the stepped leader is extended towards the ground in discrete steps dt.

Electrode gap. The gap distance H_S represents the distance between the negative leader and the structure on the ground at the inception of the connecting upward leader. The gap distance decreases with negative leader propagation (figure 12.2)

$$H_S(t + \mathrm{d}t) = H_S(t) - v_d \mathrm{d}t \tag{12.1}$$

or

$$H_S(t) = \mathrm{d} \sin \alpha \tag{12.2}$$

with

$$\tan \alpha = \frac{\mathrm{d}}{(D - h - L_z)} \tag{12.3}$$

and

$$L_Z(t + \mathrm{d}t) = L_Z(t) + v_\mathrm{d} \mathrm{d}t. \tag{12.4}$$

v_d is the downward leader velocity

Applied voltage. The applied voltage is bi-exponential and derived from the electric field changes on the ground radiated by the downward stepping negative leader. The voltage wave characteristic has been previously modelled by Lalande [3]:

$$U(t) = U_\mathrm{m}\left(1 - \mathrm{e}^{\frac{t}{t_\mathrm{m}}}\right) \text{ with } U_\mathrm{m} = E_\mathrm{m} H_S, \tag{12.5}$$

where U_{m} and E_{m} are the crest voltage and crest electric field, respectively, enhanced at the grounded object level, and

$$t_{\mathrm{m}} = \frac{30\pi\varepsilon_0(D - h)}{0.9V_{\mathrm{d}}} \quad \frac{900 I_{\mathrm{RS}}^{1/3}}{\left(8 + \dfrac{2(D - h)}{10 I_{\mathrm{RS}}^{2/3}}\right)}(E_{\mathrm{m}} - E_{\mathrm{s}}). \tag{12.6}$$

E_{s} and I are, respectively, the threshold electric field enabling the inception of an upward leader from a grounded structure and the presumed return stroke current. The current measurements at the bottom of the grounded wire indicate that a positive upward leader starts at the time the electric field at ground reaches 10 kV m^{-1} [3]. At each time step, the applied voltage is re-calculated.

Corona region. The inception conditions for the first corona (minimum threshold voltage and statistical time lag) are tested until they are satisfied. After the formation of the first corona, the leader inception conditions (i.e. the electric field criteria proposed by Dellara [3]) are tested until they are satisfactory. Then the development of the second corona begins. Following the formation of the second corona, the propagation of the continuous streamer–leader system is initiated.

Streamer–leader system. The electric-circuit based streamer–leader model is described in chapter 7. The propagation procedure together with the stop and go conditions are also described. Note that the leader is considered to be in non-LTE.

Final jump and arc phase. The transition from the free development phase to the final jump phase starts when the mean field between the negative downward leader tip and the positive upward leader reaches 500 kV m^{-1} [3]. The initiation of this phase leads inevitably to breakdown unless the applied voltage drops substantially. In the arc phase, LTE applies in the channel. The recombination of particles is the dominant process. The return stroke is characterized by a high current through the discharge channel. The value of this current is entirely defined by the characteristics of the test circuit.

The leader–return stroke electric field. The electric field E radiated by the discharge is the result of the electric fields radiated by the negative downward leader and the positive upward leader when it is initiated, and the electric field radiated by the return stroke.

Concerning the electromagnetic field effects at close distances to the lightning channel, we use the simplified model proposed by several investigators [1, 8, 9] assuming the Earth as a perfect conductor, as explained in section 10.3.

For the positive upward leader, the same approximation has been used; in this case, the velocity and charge density are time-dependent.

For the distance ranges considered, it was assumed that the return stroke fields can be calculated using the same simple approach used for the calculation of the leader electric field changes. Such an approximation has been used and validated by Rachidi *et al* [9]. The electric field changes associated with the return stroke may be computed using the relationship described in section 10.3, keeping ΔE_z^{s} constant and equal to its value at the end of the leader propagation. During the return stroke phase, the height H_{s} between the downward leader tip and the grounded structure becomes a dynamic parameter such as $H_{\mathrm{s}} = v_{\mathrm{RS}}\mathrm{d}t$.

12.2.3 Results and discussion

For the validation of the proposed model, some of the results were compared to those obtained experimentally during a triggered lightning campaign performed at Camp Blanding in the summer of 1995 [10].

The altitude triggering technique was used to study the bi-directional leader. A small rocket was launched that spooled out 400 m of wire followed by a triggering copper wire. The triggering wire was made of Kevlar-coated copper 0.2 mm in diameter [3]. When this wire was been trailed out over a sufficient length, a bi-directional leader discharge developed from its two ends. An upward positive leader was initiated from the tip of the rocket, and propagated toward the cloud. A few milliseconds later, a downward negative stepped leader developed from the bottom of the triggering wire and propagated toward the ground.

This rocket-triggered lightning technique has been widely used to study both the physics of lightning discharges and their effects on various systems [2, 3–6, 11, 12]. Rocket-triggered lightning experiments at Camp Blanding, Florida, have been carried out since 1993. Information about the Camp Blanding lightning research and testing facilities can be found in [13]. The classical lightning-triggering techniques involve the use of a small rocket extending a thin grounded wire upward.

A self-consistent model was used to simulate the connecting positive leader from the 50 m grounded wire that is supposed to represent a tall structure on the ground. The cloud–ground distance was taken to be 2000 m.

The result of the E field best fit is illustrated in figure 12.3. The electrostatic simulation is in a good agreement with the measurements.

The inception electric field threshold for the upward connecting leader was reached after 4.19 ms, which corresponds well to the beginning of the significant pulse currents recorded during the triggering experience (figure 12.4(a)). The comparison of the

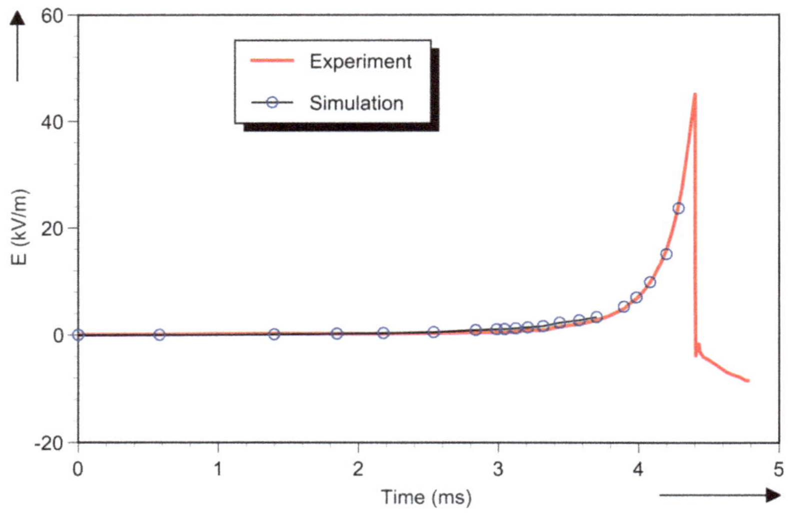

Figure 12.3. Time evolution of electric field at the ground at 50 m of the lightning channel until the beginning of the final jump phase. Comparison between measured and calculated E field.

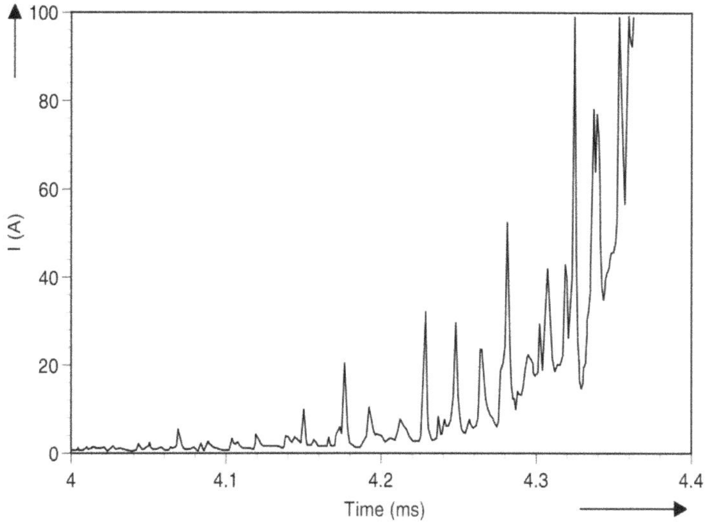

(a)

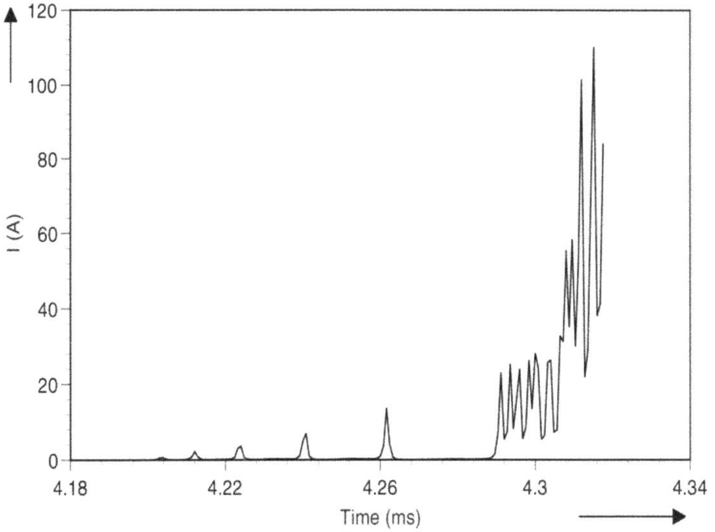

(b)

Figure 12.4. (a) Measured and (b) simulated current produced by the upward connecting positive leader from the grounded 50 m long wire, reproduced with permission from [10].

results given in figures 12.4(a) and 11.4(b) shows a good agreement between the measured and simulated currents flowing through the leader channel. This agreement concerns not only the shape but also the amplitude. The positive leader current exhibits pulses separated by approximately 20 μs. The pulse crest value increases up to 100 A—a few microseconds before the return stroke—while the electric field reaches a few tens of kV m^{-1} at ground.

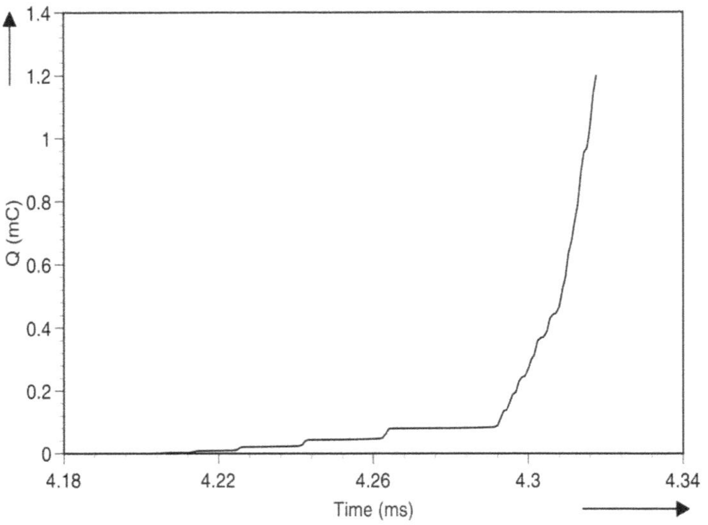

Figure 12.5. Simulated charge injected by the upward connecting positive leader from the grounded 50 m long wire.

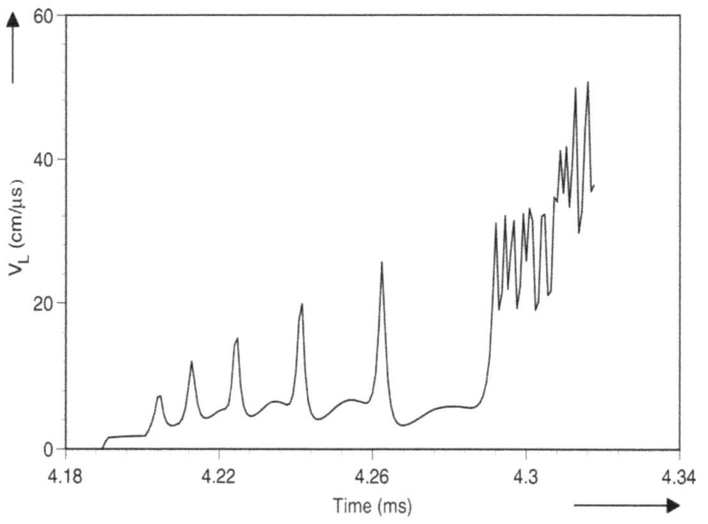

Figure 12.6. Simulated values of the upward connecting positive leader instantaneous velocity from the grounded 50 m long wire.

The simulation of the charge injected into the gap during the development of the positive upward leader is given by figure 12.5, while figure 12.6 shows the instantaneous velocity of the upward connecting leader. Figure 12.7 shows the simulation of the attachment process as could be observed with a camera working in streak mode.

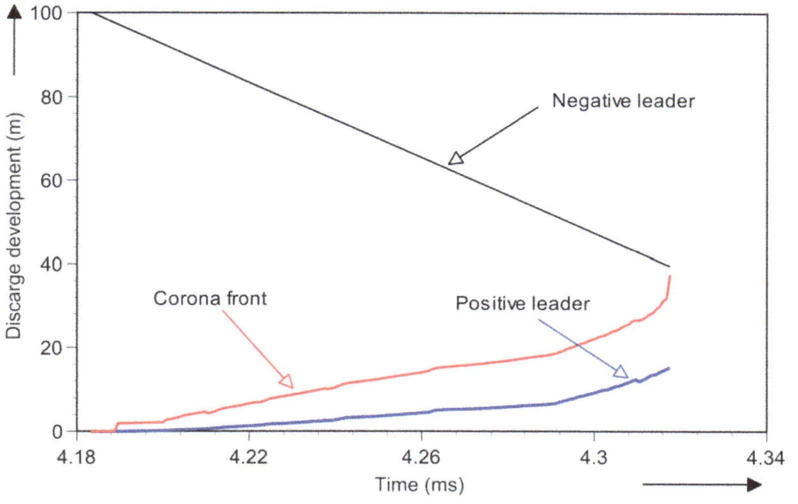

Figure 12.7. Computing of the streaked image sketch of the attachment process of the artificially triggered lightning sequence.

In atmospheric discharges, the external ambient field (50 kV m^{-1}) is relatively lower than that of the laboratory (one order of magnitude); consequently, the field at the corona front is mainly produced by the leader tip. When the leader corona-streamers develop, they propagate at a very high speed and extend rapidly away from the leader tip [4]. The relationship (12.7) allowing the prediction of the streamer length for long laboratory sparks did not provide results in accordance with those measured by Lalande *et al* [3, 4]. The results are better simulated when using the empirical formula established by Hutzler and Hutzler [14]. Based on experimental studies, these authors proposed the following relationship for the length S of streamers

$$S = \frac{P}{P_0}\frac{4.25}{1 + \dfrac{3.5}{L}} + S_i \exp\left(\frac{-L_z}{0.8S_i}\right) \tag{12.7}$$

where L_z and S_i are the leader axial length and the length of the streamers of the first corona, respectively ($S_i = 1.44U_i$ [15], U_i being the inception voltage), while P and P_0 represent the voltage at the leader tip and the voltage one would have to apply to the unabridged gap to make the field on the leader tip equal to the critical field.

At the junction, the positive leader is about 15 m long, while the corona front is about 25 m long. These values are in good agreement with the data deduced from camera recordings [3, 4].

12.3 A quantitative study of lightning striking distance factors

In this section, the results of a sensitivity analysis of an upward leader initiation from a rod in response to an approaching lightning downward leader are presented. Whilst the simple electro-geometric model (EGM) [10] provided the first quantitative measure of lightning attachment to a grounded object in terms of the prospective peak current of the lightning return stroke, I_{RS}, the model proposed hereafter includes more parameters than just I_{RS}. For example, it takes into account the height of the structure, the electric field intensification, the leader initiation and propagation conditions, and the downward leader channel characteristics.

12.3.1 Overview of striking distance calculations

The EGM has analytical tools for the study of lightning strikes on objects such as towers and overhead line conductors. One of the main applications of the EGM is, thus, the protection of power lines or sensitive installations (gas and petroleum industries, etc) against direct flash. The EGM is also used to determine the positions and lengths of lightning rods for protecting buildings against direct strikes.

The EGM was first introduced by Whitehead in the late 1960s [16, 17] and is based on field studies of natural lightning. Two important parameters of the lightning stroke to a grounded structure are the *prospective peak stroke current*, I_{RS}, and the *striking distance*, d_S. The striking distance d_S (figure 12.8) is such that if the head of the downward lightning leader falls within the striking distance of an object, the object will collect the stroke:

$$d_S = aI_{RS}{}^{b}, \qquad (12.8)$$

where d_S is in metres and I_{RS} is the prospective return stroke peak current in kA. Importantly, the simple EGM assumes that, for a given leader charge or current, the striking distance is a constant value.

The classical models are all based on the propagation of the negative stepped leader and the field that the leader induced on the ground structures. The different values of a and b (table 12.1) result from the assumption made for the charge distribution along the stepped leader [17].

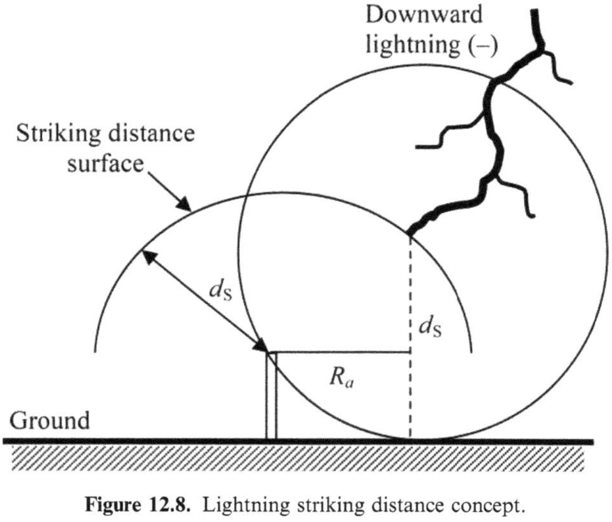

Figure 12.8. Lightning striking distance concept.

Table 12.1. Parameters *a* and *b* for some classical models.

Model	*a*	*b*
Whitehead and Brown	6	0.8
Whitehead	6.4	0.75
IEEE 1993	8	0.65
Love	10	0.65

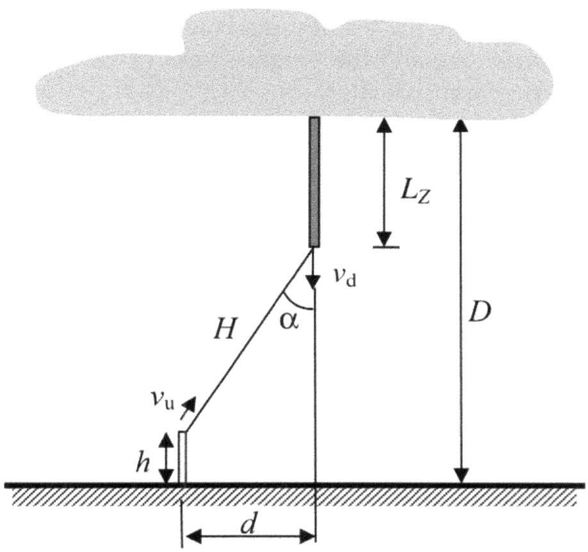

Figure 12.9. The cloud base is simulated by an infinitely large flat plate at a given potential located at a height *D* above the ground surface. This arrangement provides a uniform field between the cloud and the ground. The stepped leader channel is simulated by a perfect conductor at a cloud potential that extends in steps towards the ground.

12.3.2 General description of the modelling

The general concept of the model is presented in figure 12.9. The cloud base is simulated by an infinitely large flat surface at a given potential located at a height *D* (km) above the ground. This arrangement provides a uniform electric field between the cloud and the ground. The rod is modeled as a cylindrical conductor of height *h*, which ends in a half-sphere having a radius *r* equal to that of the cylindrical section.

The negative downward stepped leader

The physical process that takes place from the inception of a connecting leader to its successful connection to the downward stepped leader is dynamic. For the sake of simplicity, the downward negative leader is simulated by a simplified electrostatic model. The stepped leader channel is simulated by a perfect conductor, regarded as a uniformly charged line. The charge per unit length q (C m^{-1}) can be evaluated from the prospective return stroke current I_{RS} (A)

$$q = 0.43 \times 10^{-6} I_{RS}^{2/3}. \tag{12.9}$$

An approximate relationship between the total charge, Q_T (C), transferred to the ground and I_{RS} (kA) has been proposed in [18]

$$I_{RS} = 10.6 Q_T^{0.7}. \tag{12.10}$$

According to Petrov *et al* [19], the relationship between the prospective peak current of the lightning return stroke and the progress velocity of the negative downward leader may be expressed approximately as

$$v_d \cong 8 \times 10^3 I_{RS}^{1/3}. \tag{12.11}$$

The negative leader initially propagates without the influence of the positive leader. Its progression enhances the electric field on the ground.

The discharge further develops at the start of the final jump, as was already described in section 10.4.

The positive upward leader

The assumptions as well as the relationships giving the gap distance between the negative leader and the structure on the ground at the inception of the connecting upward leader, the applied voltage, the final jump and arc, and the leader–return stroke electric field are similar to those presented in section 10.4.

Concerning the leader–streamer system, the *striking distance* is basically defined as the distance between the object to be struck and the tip of the descending leader at the time the upward connecting leader is initiated from the object. The electric-circuit based leader–streamer model is described in chapter 7. The propagation procedure together with the stop and go conditions are also described.

The inception conditions for the first corona (minimum threshold voltage and statistical time lag) are tested until they are satisfied. Typically, this involves an analysis of the electric field at the grounded structure top due to the approach of the downward leader. This field, influenced by the dynamic field of the approaching downward leader, is evaluated analytically taking into account the 'electric field intensification factor' K_i, of a grounded structure of height H. The value K_i is the ratio of the intensified electric field at the structure top E_P to the value E_0 of the ambient field.

$$K_i = E_P / E_0. \tag{12.12}$$

Protrusions or whiskers are known to cause a strong enhancement of the local field, as shown in figure 12.10. A field enhancement factor $K_i \approx 2h/w$ is defined as $E_0 = K_i E$, where E_0 is the local field strength at the top of the protrusion and E is the unaffected field strength further down in the gap [20].

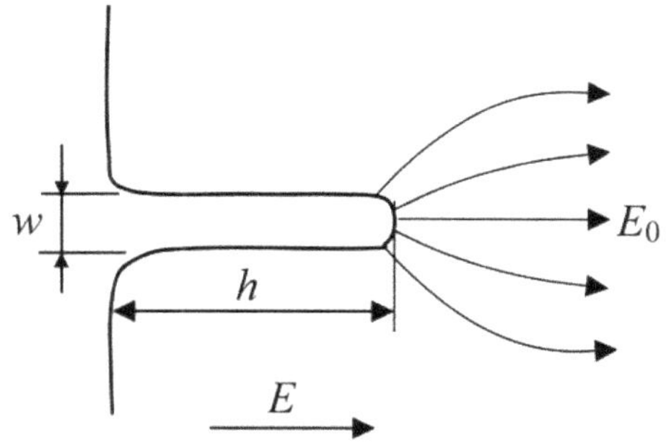

Figure 12.10. Field enhancement caused by a sharp protrusion at an electrode in a vacuum. At the top of the protrusion the field is β times higher than the average field.

In each step, an investigation is made to check whether the conditions necessary for the streamer inception from the rod, or streamer–leader transition have been reached. This point is easily evaluated using Peek's law (see equation (3.3)).

After the formation of the first corona, the leader inception conditions (i.e. the electric field criteria proposed by Hutzler [21]) are tested until satisfactory. Then the development of the second corona starts. Following the formation of the second corona, the propagation of the continuous streamer–leader system is initiated.

The electric field strength for air breakdown that is used in the application of the leader inception criterion is corrected for variations in altitude z (in km) and air pressure P (in kPa).

The application of such a correction is particularly important when considering lightning protection systems installed at high altitudes [18]. A variation is modelled using a simplified linear approximation [18]:

$$E(z) = E_0(1–0.09z), \tag{12.13}$$

Where $E_0 = 3$ MV m^{-1} represents the breakdown field strength at atmospheric pressure and sea level [18]. Changes of pressure with altitude can be approximated using

$$P(z) = P_0 \exp(-z/8), \tag{12.14}$$

where P_0 is the standard atmospheric pressure taken at 1.013 kPa and z is the altitude (km) [18].

Humidity corrections change E_0 by up to 5% and hence can be ignored [18]. Recall that humidity variations are governed almost entirely by the prevailing weather conditions during the thunderstorm.

12.3.3 Results and discussion

Figure 12.11 shows the electric field changes on a rod of height h =15 m at different lateral positions, d, from the downward lightning channel. The cloud–ground distance was taken as 5000 m with a downward leader mean velocity of 0.85×10^7 m s^{-1} and charge per unit length $q = 75$ μC m^{-1}.

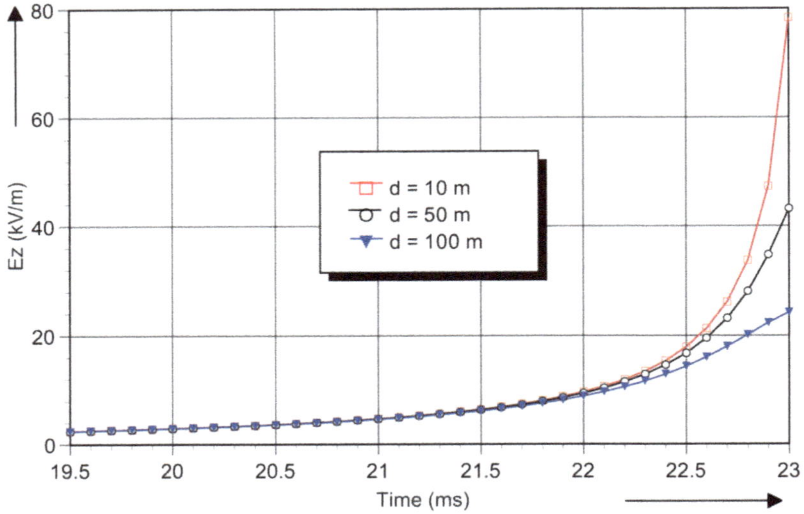

Figure 12.11. Electric field changes from different lateral positions of a grounded structure of height $h = 15$ m and radius $r = 0.1$ m with parameters $q = 0.296$ mC^{-1}, $I_{RS} = 20$ kA, and $v_d = 2.15 \times 10^5$ m s^{-1} for cloud–ground distance $H = 5000$ m.

Table 12.2. Charge-current equivalence for negative lightning.

I_{RS} (kA)	% strikes > I_{RS}	q (mC m^{-1})	Q_T (C)	$d_{S,EGM}$ (m)
5	99	0.13	0.6	28.5
10	92	0.2	0.97	44.7
20	65	0.31	1.56	70.1
25	50	0.26	1.83	81.0
50	20	0.57	2.96	127.2
75	10	0.74	3.93	165.5
100	4	0.9	4.89	199.5

Table 12.2 gives some examples of the charge-current equivalences calculated from (12.9) and (12.10), taking into account the standard cumulative frequency distribution of the peaks' currents.

The striking distances, $d_{S,EGM}$, calculated using the simple EGM (12.8) and Love's parameter from table 12.1 are listed in table 12.2 for comparison purposes.

Figure 12.12 shows the striking distance as a function of rod height for two different radii of curvature of the rod electrode. The cloud–ground distance assumed in all calculations was 5000 m. Rods of a practical height ($1 < h < 15$ m) and tip radius of curvature ($0.1 < r < 0.2$ m) on the ground surface were considered.

Out of this figure, it can be seen that the height and tip radius of the grounded rod are significant factors affecting the striking distance.

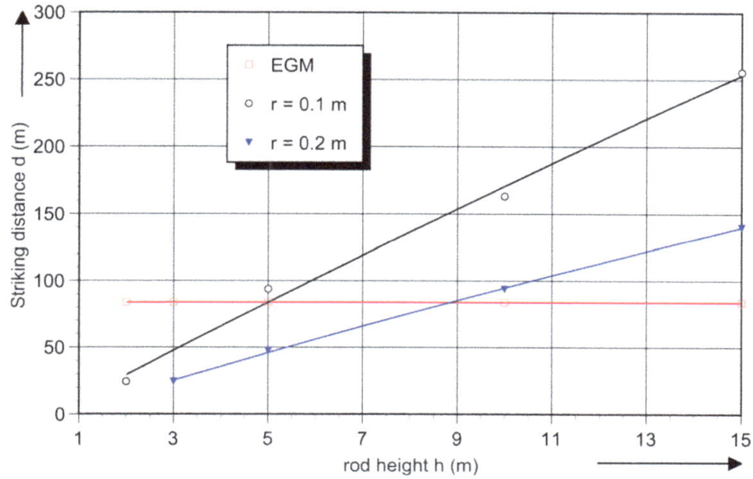

Figure 12.12. Striking distance as a function of rod height and tip radius of curvature, *r*. The peak return stroke current is fixed at 25 kA.

The simple EGM provided the first quantitative measure of lightning attachment to a grounded structure in terms of the return stroke peak current [17]. However, in existing lightning protection design methods, there is no quantitative method accounting for the: (i) main lightning-related parameters, such as the leader charge and its distribution, leader propagation, cumulative frequency distribution of lightning stroke currents, etc; (ii) the effects of altitude on air breakdown parameters and the subsequent results; (iii) the wide range of structure shapes and dimensions found in practice; (iv) the height of the air terminals used for protection; (v) the electric field intensification created by the structure and its features as a basis for upward leader inception; and (vi) the variation in the lightning attachment probability of different competing features on structures. The proposed model takes into account more parameters than just the peak current. These parameters include grounded structure tip radius and height, electric field intensification, streamer and leader initiation, propagation conditions, downward leader channel characteristics, atmospheric parameters (e.g. air density and humidity), and upward leader tortuous propagation and branching. The key parameter for striking distance determination is the electric field intensification at the rod tip; the implementation of finite element or charge simulation methods to compute the electric field distribution over and around an Earth-bound structure and its micro-geometry should therefore be considered.

References

[1] Wang D, Rakov V A, Uman M A, Takagi N, Watanabe T, Crawford D E, Rambo K J, Schnetzter G H, Fisher R J and Kawasaki Z I 1999 *J. Geophys. Res.* **104** 2143

[2] Uman M A 1987 *The Lightning Discharge* (*International Geophysics Series* vol 39) (New York: Academic)

[3] Lalande Ph 1996 Study of the lightning stroke conditions on a grounded structure *PhD Thesis* ONERA, Châtillon

[4] Lalande P, Bondiou A, Laroche P, Bachiega G L, Bonamy A, Gallimberti I, Eybert-Berard A, Berlandis J P and Bador B 1996 Modelling of the lightning connection process to a ground structure *23rd Int. Conf. Lightning Protection* (*Firenze, Italy,* 23–27 *September*) p 429
[5] Uman M A, McLain V P, Fisher R J and Krieder 1973 *J. Geophys. Res.* **78** 3530
[6] Willett J C, Idone V P, Orville R E, Leteinturier C, Eybert-Berard A, Barret L and Krieder E P 1988 *J. Geophys. Res.* **93** 3867
[7] Boudissa R, Belaicha R and Boubakeur A 1998 Experimental and theoretical study and the influence lightning conductor's form on its zone of protection *24th Int. Conf. Lightning Protection* (*Birmingham, United Kingdom,* 14–18 *September*) p 483
[8] Diendorfer G 1990 *IEEE Trans. EMC* **32** 292
[9] Rachidi F, Rubinstein M, Nucci C A and Guerrieri S 1994 Influence of the leader electric field change on voltages induced by very close lightning on overhead lines *22nd Int. Conf. Lightning Protection* (*Budapest, Hungary,* 19–23 *September*) paper R 3b-10
[10] Le Roy G, Gary C, Hutzler B, Lalot J and Dubanton C 1984 *Les Propriétés Diélectriques de l'Air et les Très Hautes Tensions* (*Collection de la Direction des Études et Recherches d'Électricité de France* vol 51) (Paris: Eyrolles)
[11] Yokoyama S, Miyake K, Suzuki T and Kanao S 1990 *IEEE Trans. Power Deliv.* **5** 1418
[12] Wang D, Kawasaki Z I, Yamamoto K, Matsuura K, Chang J S and Janischewskyj W 1995 *J. Geosphys. Res.* **100** 11661
[13] Rakov V A and Uman M A 2005 *Lightning: Physics and Effects* (Cambridge: Cambridge University Press)
[14] Nucci C A, Rachidi F, Ianoz M V and Mazzetti C 1993 *IEEE. Trans. EMC* **35** 75
[15] Fofana I and Beroual A 1997 *J. Phys. D: Appl. Phys.* **30** 1653
[16] Golde R H (ed) 1977 *Lightning. Vol 2: Lightning Protection* (New York: Wiley)
[17] Sabot A 1995 An engineering review on lightning, transients over-voltages and the associated elements of electromagnetic compatibility *9th Int. Symp. on High Voltage Engineering* (*Graz, Austria, August* 28–*September* 1) paper 9006
[18] D'Alessandro F and Gumley J R 2001 *J. Electrostatics* **50** 279
[19] Petrov N I and Waters R T 1995 *Proc. R. Soc. Lond.* A **450** 589
[20] Kreuger F H 1991 *Industrial High Voltage* (Delft: Delft University Press)
[21] Hutzler B and Hutzler D 1992 *Bull. Dir. Etud. Rech.* **4** 11

Chapter 13

Application of the model to evaluate the induced effects on overhead lines due to a nearby positive lightning downward leader

13.1 Introduction

Transients caused by lightning are one of the major causes of malfunction or even destruction of electrical systems. During the discharge, the most energetic and most dangerous phase is the return stroke. However, it has been shown [1, 2] that at short distances from the discharge channel, the magnitude of the field radiated by the leader is the same as that of the return stroke. Indeed, near the discharge the electromagnetic fields are marked by an electrostatic aspect. Hence the leader shows radiated fields of high strength.

Due to their metallic nature, overhead lines behave as receivers for the electromagnetic field radiated by the atmospheric discharges. Evaluating the risks of overvoltages of atmospheric origin occurring on overhead power supply lines is useful for improving the co-ordination of insulation and electromagnetic compatibility.

Calculations of currents and voltages induced on a line parallel to the ground have been performed by many authors [3–11]. Most of these studies concern negative cloud–ground flashes and the return stroke electric field change, neglecting any field changes prior to it. It has been shown theoretically [2, 7] and experimentally [2, 7, 12] that at a few tens of metres from the lightning channel, the field radiated by the leader shows a magnitude close to that of the return stroke. Therefore, the peak voltages induced on distribution lines by the leader, preceding a return stroke, may impose an additional stress on lightning protection devices to those brought by the return stroke [7].

The mathematical model proposed in chapter 11 to satisfactorily describe a positive lightning stepped leader is used to evaluate the electromagnetic pulses associated with a nearby positive lightning downward leader. The transient voltages and currents induced on a nearby line are evaluated by assuming different strike

doi:10.1088/978-0-7503-1236-3ch13

points in the vicinity of the line. This model allows for the determination, at each step, of the charge distribution in the leader, in contrast to those studies performed on lightning leader induced over-voltages which assume a constant value through the propagation.

This chapter presents simulation results for specific engineering applications, using the mathematical model proposed in chapters 7 and 11, examining the induced effects on an overhead line due to a nearby positive lightning downward leader without and in the presence of a distribution transformer and a surge arrester.

13.2 Induced effects on an overhead line due to nearby positive lightning downward leader

13.2.1 The vertical electric field associated with the leader

The vertical electric field at distances less than a few kilometres from the lightning channel can be calculated with reasonable approximation, assuming the ground is a perfect conductor [3–9].

The leader electric field at distances close to the lightning channel can be determined using the simplified diagram in figure 13.1, where the ground is considered as a perfect conductor. The leader channel is represented by successive straight segments charged from an initial charge source situated at the cloud base.

The electric field at $M(x_0, y_0, z_0)$ due to each element of charge $\rho(z)\mathrm{d}z$ can be expressed as [1, 2, 13]

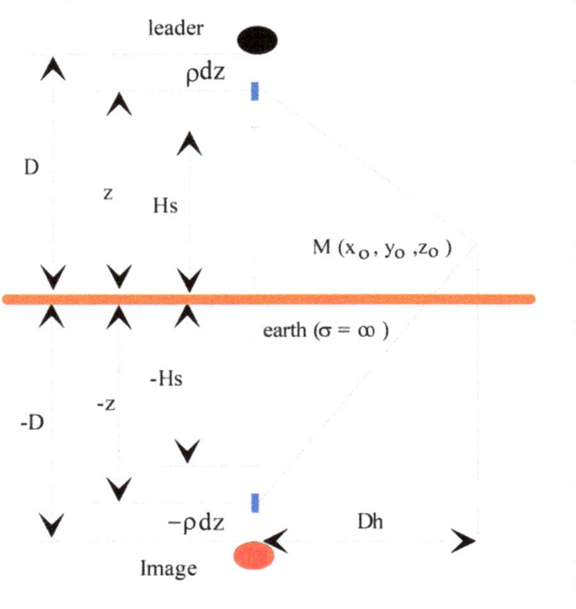

Figure 13.1. Simplified geometric model used for the leader electric field calculation.

$$dE(\rho dz) = \frac{\rho(z)dz}{4\pi\varepsilon_0\left((z - z_0)^2 + Dh^2\right)}, \tag{13.1}$$

with $r = Dh = \sqrt{x_0{}^2 + y_0{}^2}$ and the linear distribution of the charge along the leader channel.

The vertical electric field due to the actual length of the leader is:

$$E_{ZL}(t, r, z) = \frac{1}{4\pi\varepsilon_0}\int_{H_s}^{D} \frac{\rho(z)(z - z_0)dz}{\left((z - z_0)^2 + Dh^2\right)^{3/2}}. \tag{13.2}$$

The decrease of charges in the initial source located at the cloud base affects the electric field change at $M(x_0, y_0, z_0)$. Assuming that the charges in the source have a spherical distribution, the variation of the electric field due to this decrease in charge at time t is:

$$E_z^{\,s} = -\frac{\rho(z)}{4\pi\varepsilon_0}\frac{(D - H_s)(D - z_0)}{Dh^3\left(1 + (D - z_0)^2/Dh^2\right)^{\frac{3}{2}}}. \tag{13.3}$$

The total vertical electric field at M above a perfectly conducting ground can be expressed as the sum of:

- the field radiated by the leader and the contribution of charge reduction in the initial source and
- the contribution of their images with respect to the Earth.

The expression of it at time t is given by the following relationship:

$$E_z(t, r, z) = \frac{1}{4\pi\varepsilon_0}\left[\int_{H_s}^{D} \frac{\rho(z)(z - z_0)dz}{\left((z - z_0)^2 + Dh^2\right)^{3/2}} - \int_{-H_s}^{-D} \frac{\rho(z)(z + z_0)dz}{\left((z + z_0)^2 + Dh^2\right)^{3/2}}\right] + \Delta E_z^{\,s}(t, r, z) \tag{13.4}$$

with

$$\Delta E_z^{\,s}(t, r, z) = -\frac{\rho(z)}{4\pi\varepsilon_0}\left[\frac{(D - z_0)}{\left((D - z_0)^2 + Dh^2\right)^{3/2}} + \frac{(D + z_0)}{\left((D + z_0)^2 + Dh^2\right)^{3/2}}\right](D - H_s), \tag{13.5}$$

where $\rho(z)$ represents the linear distribution of the charge along the leader and $H_s = D - L_z$.

13.2.2 The horizontal field associated with the leader

The horizontal electric field plays an important role in the calculation of induced voltage on overhead lines. Its calculation requires the use of rigorous expressions or at least reasonable approximations. In the literature [3–11], three main approaches have been proposed to calculate the horizontal electric field, namely: (i) the perfectly conducting ground approximation, (ii) the wave-tilt formula, and (iii) the Cooray–Rubinstein approach.

Perfect the conducting ground approximation
Assuming the ground is a perfect conductor, the horizontal field radiated by the leader channel at $M(x_0, y_0, z_0)$ is given by the following expression in the time domain

$$E_r(t, r, z) = \frac{Dh}{4\pi\varepsilon_0}\left[\int_{H_s}^{D} \frac{\rho(z)\mathrm{d}z}{\left((z - z_0)^2 + Dh^2\right)^{3/2}} + \int_{-H_s}^{-D} \frac{\rho(z)\mathrm{d}z}{\left((z + z_0)^2 + Dh^2\right)^{3/2}}\right], \tag{13.6}$$

where ΔE_r^s represents the effect of charge depletion in the cloud and is

$$\Delta E_r^s(t, r, z) = -\frac{\rho(z)(D - H_s)Dh}{4\pi\varepsilon_0}\left[\frac{1}{\left((D - z_0)^2 + Dh^2\right)^{3/2}} + \frac{1}{\left((D + z_0)^2 + Dh^2\right)^{3/2}}\right]. \tag{13.7}$$

This approximation has been used in some other studies in the evaluation of induced voltages on transmission lines due to nearby lightning [5–7, 9]. At short distances (Dh = 30–200 m) and for ground conductivities of about 10^{-2} S m^{-1} or higher, the perfect conducting ground assumption can be considered reasonable for an observation point located a few metres above the ground [9].

Wave-tilt formula
The wave-tilt formula allows one to obtain the horizontal component of the electric field over a ground of finite conductivity, provided the vertical component is known:

$$W(j\omega) = \frac{E_r(j\omega)}{E_z(j\omega)} = \frac{1}{\sqrt{\varepsilon_{rg} + \sigma_g/j\omega\varepsilon_0}}, \tag{13.8}$$

where σ_g, ε_{rg} and ε_0 are the ground conductivity, the relative permittivity, and the permittivity of free space, respectively. This approach is valid for radiated fields with a grazing angle of incidence with respect to the ground [5, 6, 9]. The use of this function for the case of lightning is therefore reasonable only at far observation points [14, 9].

Cooray–Rubinstein approach
The Cooray–Rubinstein approach was proposed to calculate the horizontal electric field above a finitely conducting ground [9]. The electric field is viewed as being decomposed into two terms. The first one is the horizontal field calculated assuming

the ground to be perfectly conducting. The second is determined from the azimutal magnetic field using the following surface impedance expression:

$$E_r(r, z, j\omega) = E_{rp}(r, z, j\omega) - H_{\varphi p}(r, 0, j\omega)\frac{c\mu_0}{\sqrt{\varepsilon_{rg} + \sigma_g/j\omega\varepsilon_0}}, \qquad (13.9)$$

where c, μ_0, are the speed of light in free space and the permeability of free space, respectively. The Cooray–Rubinstein approach permits us to obtain satisfactory results for all the considered ranges [9].

13.2.3 Field–transmission line coupling equations in the time domain

From the range of distances considered ($r = Dh = 50$–100 m), the horizontal electric field may be calculated using a perfectly conducting ground approximation, which greatly reduces the computation time [9, 15]. It is possible to establish different forms of the equations describing the coupling of an external electromagnetic field applied to an overhead line of length L_x (see figure 13.2) on the basis of Maxwell's equations along with transmission line approximations. The equations can be solved either in

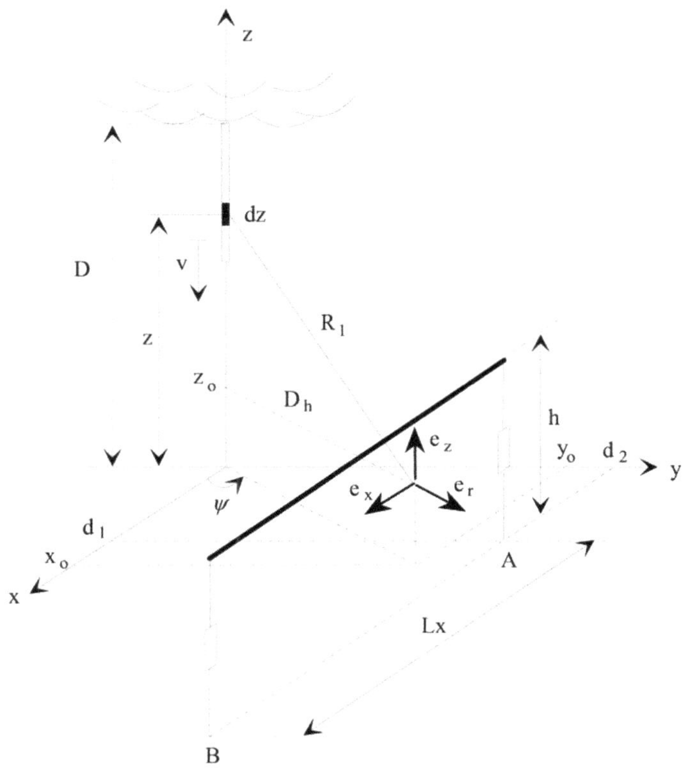

Figure 13.2. Geometry used for the calculation of voltages induced by a downward leader on a nearby overhead line.

the time or frequency domain. Writing coupling equations in the time domain allows a much more direct solution of transient problems and also facilitates the non-linearity considerations. However, the coupling equations in the time domain can also be written easily if some parameters per unit length (e.g. surface admittance of the conductor) are considered to be independent of frequency. It has been verified that such an approximation is quite valid within the framework of this study.

It has been shown [3–11, 16–18] that among the various models coupling to a horizontal conductor adopted in electrical power, only the model by Agrawal *et al* [19] and its equivalent formulation [11] are to be considered rigorous within the limits of the transmission line approximation. According to Agrawal *et al*, coupling equations are written in terms of a total current on the line and scattered voltage

$$\partial U^{s}(x, t) + R' i(x, t) + L' \frac{\partial i(x, t)}{\partial t} = -E_z^i(x, h, t) \tag{13.10}$$

$$\frac{\partial i(x, t)}{\partial x} + G' U^{s}(x, t) + C' \frac{\partial U^{s}(x, t)}{\partial t} = 0, \tag{13.11}$$

where $E_x^{\,i}(x, h, t)$ represents the horizontal components of the incident electric field along the x-axis. Furthermore, $i(x, t)$ is the current induced along the line; R', L', G', and C' are the parameters (resistance, inductance, conductance, and capacitance, respectively) per unit length of the line.

The total induced voltage $U(x, t)$ is given by the sum of the scattered voltage $U^{s}(x, t)$ and the so-called incident voltage $U^{i}(x, t)$

$$U^{i}(x, t) = -\int_0^h E_z^i(x, z, t)\mathrm{d}z = -E_z^i(x, 0, t)h, \tag{13.12}$$

where $E_z^{\,i}$ represents the vertical electric field.

The line is terminated at both ends by resistances of about 500 Ω (which may be considered the characteristic impedance of a lossless line).

The boundary conditions (figure 13.2) taking into account the coupling of the vertical electric field at the two terminations are as follows:

$$U^{s}(0, t) = -R_0 i(0, t) - U^{i}(0, t) \tag{13.13}$$

$$U^{s}(L_x, t) = -R_0 i(L_x, t) - U^{i}(L_x, t). \tag{13.14}$$

The equivalent coupling circuit describing the equations (13.10)–(13.14) is represented in figure 13.3.

L' is the external per unit length inductance calculated for a lossless wire above a perfectly conducting ground [8, 9]

$$L' = \frac{\mu_0}{2\pi} \cos h^{-1}\left(\frac{h}{a}\right) \tag{13.15}$$

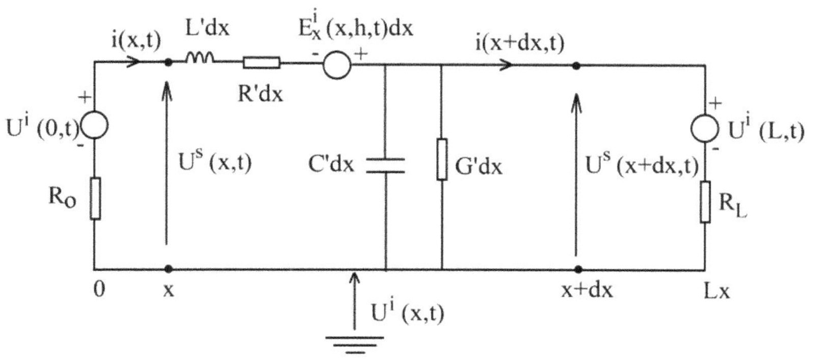

Figure 13.3. Equivalent 'coupling' circuit describing the electromagnetic coupling between an incident electromagnetic field and a transmission line [11].

for $h \gg a$:

$$L' \approx \frac{\mu_0}{2\pi} \ln\left(\frac{2h}{a}\right), \tag{13.16}$$

where h and a are the line's height and radius, respectively.

C' is the per unit length transverse capacitance calculated for a lossless wire above perfectly conducting ground, given by [8, 9]

$$C' = \frac{2\pi\varepsilon_0}{\cos h^{-1}\left(\frac{h}{a}\right)} \tag{13.17}$$

for $h \gg a$

$$C' \approx \frac{2\pi\varepsilon_0}{\ln\left(\frac{2h}{a}\right)}. \tag{13.18}$$

G' is the per unit length transverse conductance given by [8, 9]

$$G' \approx \frac{\sigma_{air}}{\varepsilon_0} C'. \tag{13.19}$$

13.2.4 Voltage and currents induced on an overhead line due to a nearby positive lightning downward leader

The simulations have been performed for a 0.5 km long, 10 m high single wire overhead line matched at both ends. The wire and the ground were assumed to be perfect conductors. Several strike locations, shown in figure 13.4, have been considered.

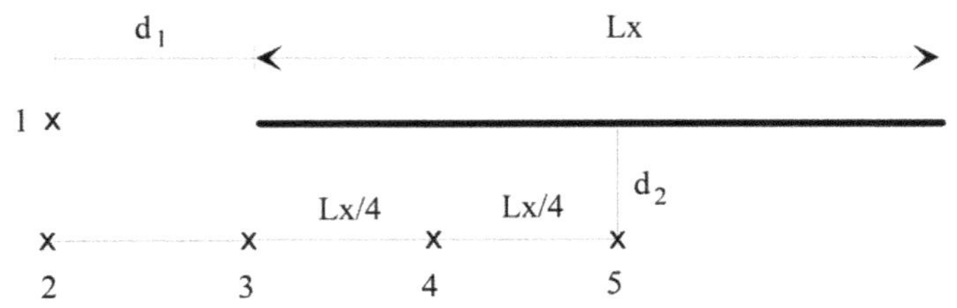

Figure 13.4. Map of considered strike point locations. The line is matched at both ends, and its height above ground is 10 m. Figure printed with permission from CIGRE 'Lightning-induced over-voltages on distribution overhead lines in presence of distribution transformers or short cables' © 1993 [20].

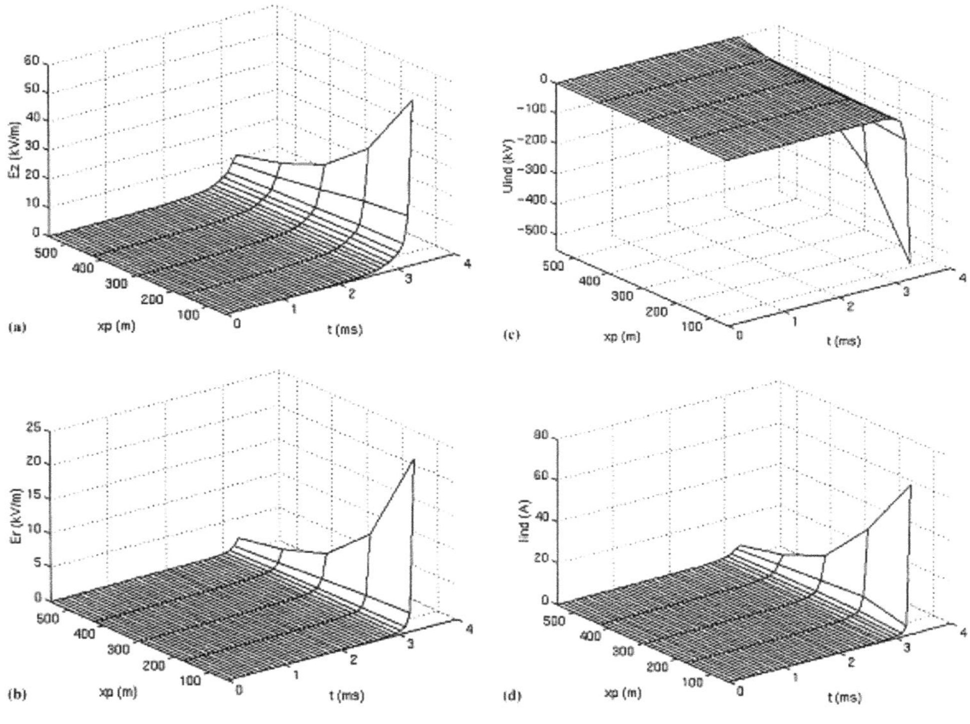

Figure 13.5. Leader electric field components and induced effects for strike location 1 in figure 13.4 where $d_1 = 50$ m. (a) Leader vertical electric field along the line at ground level. (b) Leader horizontal electric field along the line at a height of 10 m. (c) Leader induced voltage along the line. (d) Leader induced current along the line. Reprinted with permission from [21].

For each case, the horizontal and vertical field components on the one hand and induced voltage and current on the other have been computed at five equidistant points along the line. The simulation results are shown in figures 13.5–13.9. In each case, the striking location close to the line is the most affected.

In the literature, there are no experimental data for positive ground flashes giving these characteristics in given specific conditions for the same strike. However, it is known [1] that electromagnetic fields from positive flashes observed on a slow time

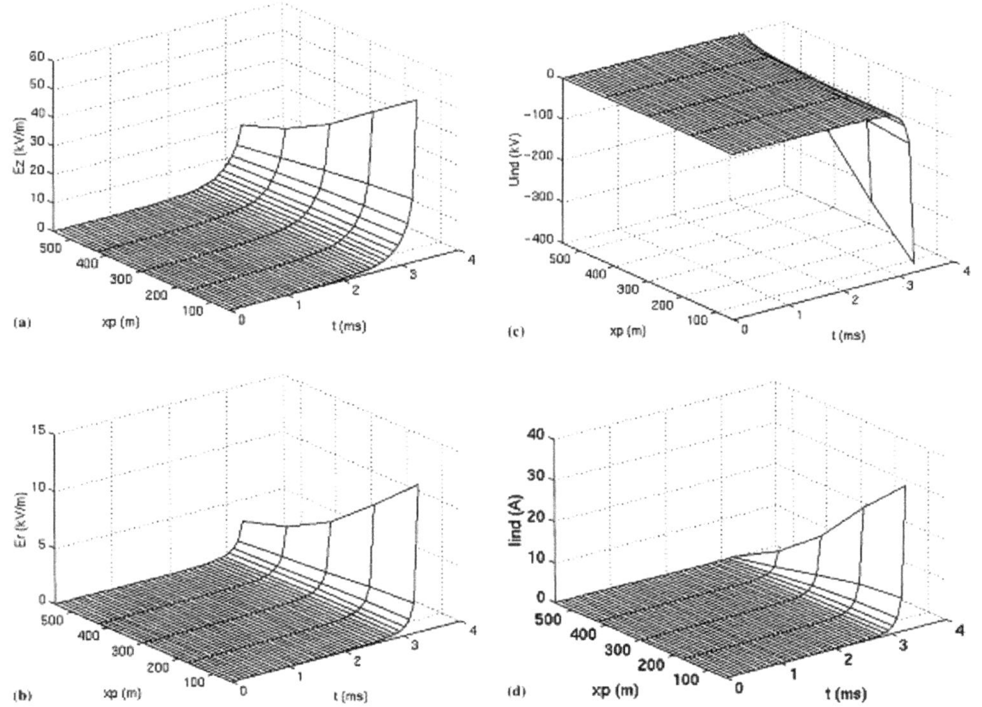

Figure 13.6. Leader electric field components and induced effects for strike location 2 in figure 13.4. (a) Leader vertical electric field along the line at ground level. (b) Leader horizontal electric field along the line at a height of 10 m. (c) Leader induced voltage along the line. (d) Leader induced current along the line. Reprinted with permission from [21].

scale are apparently similar to fields from negatives flashes, except for the polarity. It can be noted that the computed values represent a good approximation of the measured values for negative Earth flashes [1, 2, 6, 7, 12, 14, 15, 22, 23–25]. This agreement concerns not only the shape but also the amplitude of each parameter. The average peak values of the induced voltage and current are about –450 kV and 40 A, respectively.

The computed results show that, for striking locations approximately along the line prolongation (cases 1 and 2 in figures 13.4, 13.6), the induced effects are less pronounced, contrary to striking locations perpendicular to the line (figures 13.7–13.9).

From the analysis of these different figures, it appears that all parameters characterizing the positive lightning discharge given by our model are in good agreement with those recorded during natural Earth flashes.

Perhaps the most important conclusions of this study are that surge protections devices could be triggered by the leader induced voltage before the return stroke. Further studies taking into account the full lightning phases (leader and return stroke) are needed to draw more general conclusions.

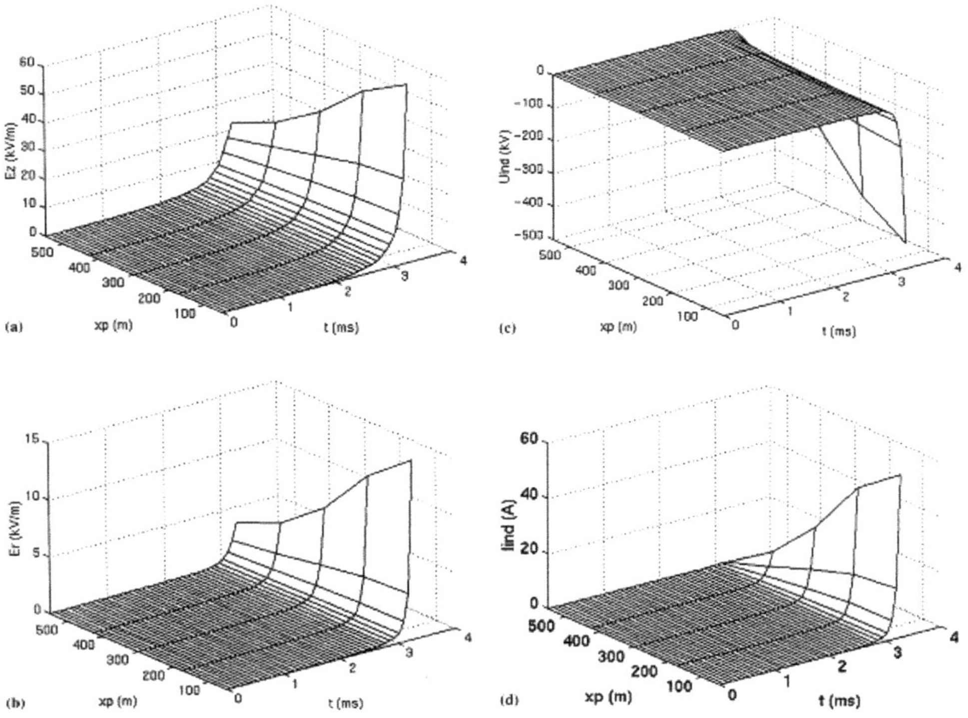

Figure 13.7. Leader electric field components and induced effects for strike location 3 in figure 13.4. (a) Leader vertical electric field along the line at ground level. (b) Leader horizontal electric field along the line at a height of 10 m. (c) Leader induced voltage along the line. (d) Leader induced current along the line. Reprinted with permission from [21].

13.2.5 Analysis of a positive lightning downward leader field coupling to a nearby overhead line located on a non-uniform ground surface

A number of studies related to lightning field coupling to transmission lines have been focussed on the effect of lightning stroke distance to the transmission line [5, 26–31], while other authors have investigated the influence of ground conductivity [32]. To the best of our knowledge, no previous study had considered lightning field changes in calculating the induced voltages on overhead lines, taking into account the non-uniformity of the ground surface.

In this section, the field coupling of a downward lightning discharge onto overhead lines situated on non-uniform/uniform ground surfaces is investigated, adopting a simple model already used by Rachidi *et al* [7]. Making use of a coupling model based on the transmission line theory for field–overhead line coupling calculation, the transient voltages induced on a line of finite length were determined, assuming different strike points in the vicinity of this line. The results obtained were compared to those obtained for a uniform ground surface and might be helpful for the proper design and coordination of the line protection and insulation since ground surfaces are generally non-uniform.

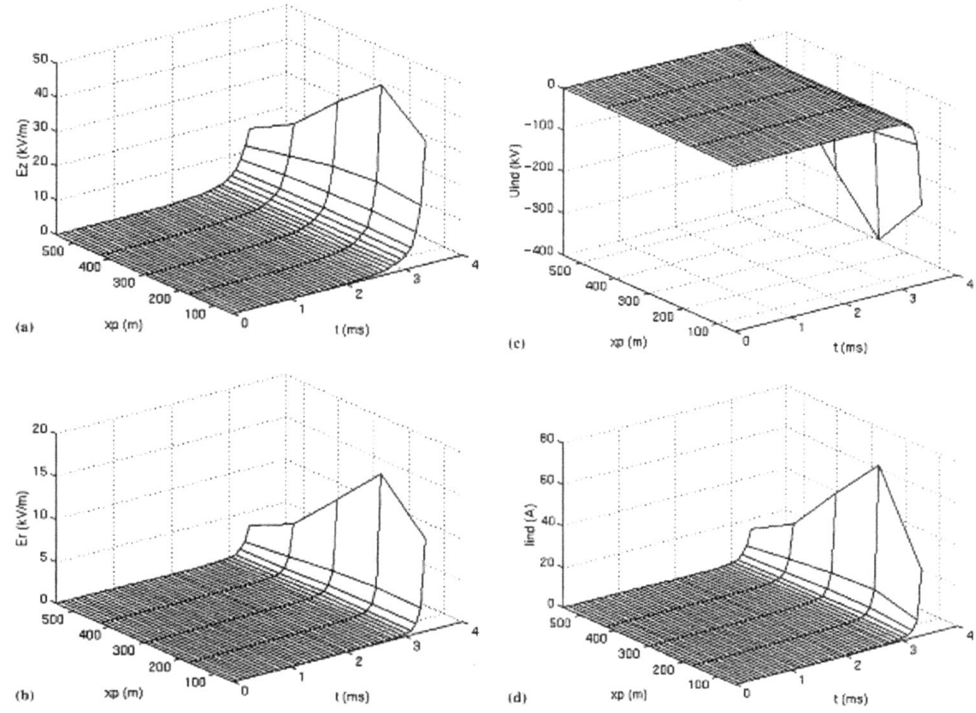

Figure 13.8. Leader electric field components and induced effects for strike location 4 in figure 13.4. (a) Leader vertical electric field along the line at ground level. (b) Leader horizontal electric field along the line at a height of 10 m. (c) Leader induced voltage along the line. (d) Leader induced current along the line. Reprinted with permission from [21].

The assumption of ground surface uniformity was dropped and the previous theories extended to non-uniform ground surface (figure 13.10(a)).

This means that the height of the line above the ground is simply a function of the position x along the line. The non-uniform ground surface is simplified as a cascaded chain of k sections of uniform ground surface. To reflect ground surface non-uniformity fully, this system is later approximated by a cascaded series of many short sections of uniform ground surface (figure 13.10). On the basis of these considerations it is essential to find piecewise solutions for segmented non-uniform ground surfaces.

For the case studied (figure 13.10), each ground surface section is uniform and the height h_k ($k = 1, 2, \ldots, 7$) is constant within each section. A cloud–ground distance D of 3 km was considered. A constant voltage $U_c = 40$ MV was applied between the Earth and the cloud considering the specific atmospheric conditions as follows: $P = 100$ kPa and $T = 25$°C at ground level. As concerns the cloud diameter, a kilometre constitutes a good dimension order [23, 32, 33]. The height of the $L_x = 500$ m transmission line matched at both ends and with a diameter of 1 cm was taken to be equal to 10 m. The analysis was performed for lightning strike locations 1, 3, and 5 (figure 13.4).

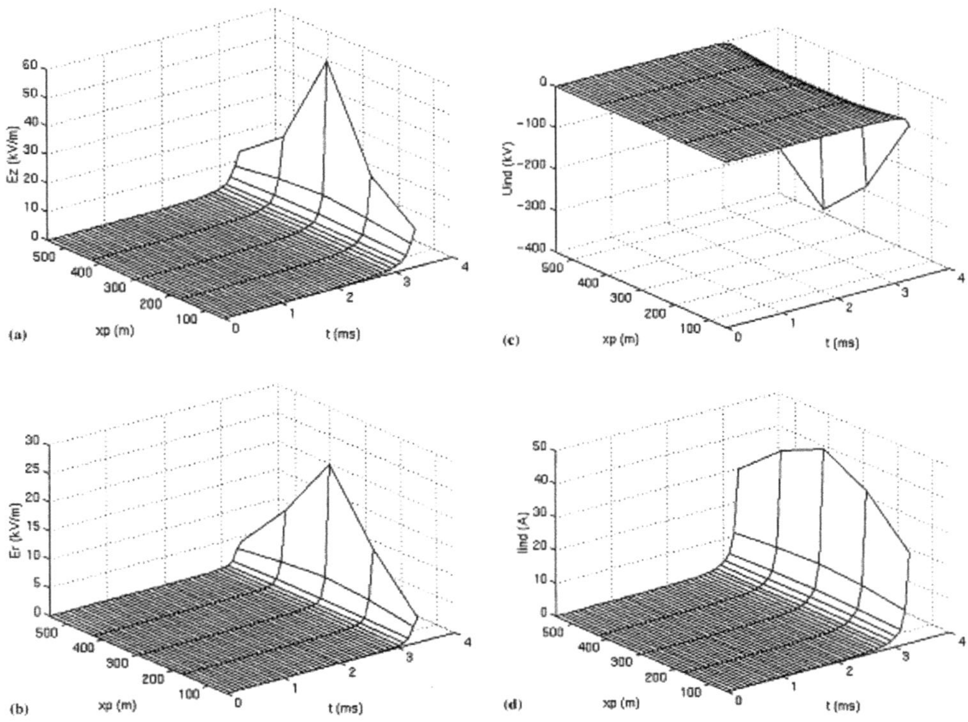

Figure 13.9. Leader electric field components and induced effects for strike location 5 in figure 13.4. (a) Leader vertical electric field along the line at ground level. (b) Leader horizontal electric field along the line at a height of 10 m. (c) Leader induced voltage along the line. (d) Leader induced current along the line. Reprinted with permission from [21].

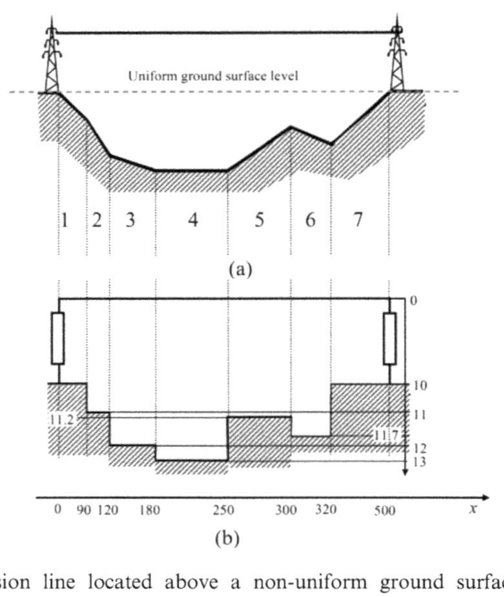

Figure 13.10. A transmission line located above a non-uniform ground surface excited by a lightning electromagnetic wave: (a) geometry and (b) piecewise uniform representation.

The assumption proposed by Diendorfer [34] was considered, whereby a nearby lightning strike is defined as one whose minimum distance to the transmission line ranges from 50–500 m. Distances of less than 50 m will probably result in a direct strike to the overhead line. Figures 13.11–13.13 represent the voltage induced on the transmission line for strike locations 1, 3, and 5 with $d_1 = d_2 = 50$ m. For each strike location the induced effects have been computed considering the transmission line located on uniform and non-uniform ground surfaces. The induced voltage has been computed at five equidistant points along the line (*xp*-axis in the figures), while for the non-uniform ground surface the positions along the *xp*-axis are related to the piecewise uniform representation (figure 13.10).

From these figures, it can be seen that the non-uniformity of the ground surface affects the induced effects. This concerns not only the shape but also the amplitude of the voltage induced along the line. For this study case (figure 13.10), the valley-like non-uniformity selected increases the height between the line and the ground. In the case of a ground surface with asperities, which would reduce the distance

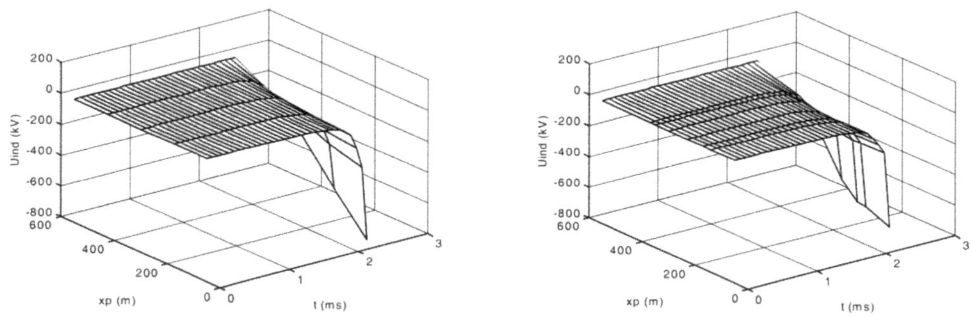

(a) (b)

Figure 13.11. Induced voltage along the line located on (a) a uniform ground surface for strike location 1 and (b) a non-uniform ground surface for strike location 1.

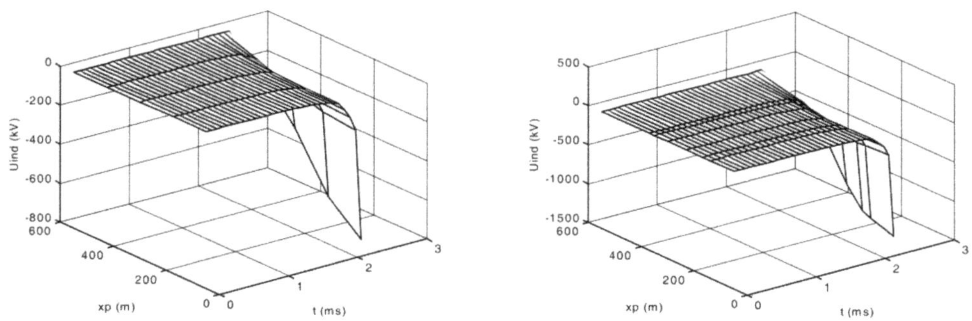

(a) (b)

Figure 13.12. Induced voltage along the line located on (a) a uniform ground surface for strike location 3 and (b) a non-uniform ground surface for strike location 3.

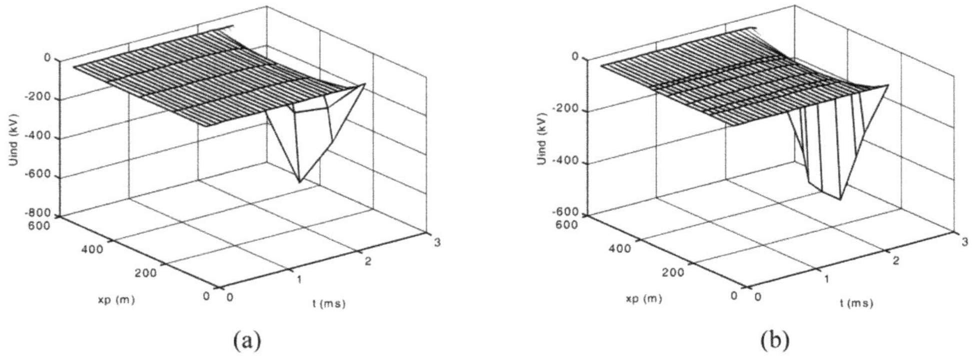

Figure 13.13. Induced voltage along the line located on (a) a uniform ground surface for strike location 5 and (b) a non-uniform ground surface for strike location 5.

between the ground and the transmission line, it would also have enhanced the induced effects.

Transmission line designers normally take ground–surface non-uniformity into consideration, since relevant standards and EPRI books [35] do recommend a minimum height (between Earth asperities and transmission lines). Precautions should, therefore, be taken to avoid height reductions. Such a reduction may result from various phenomena, such as ice/snow accumulation on the ground, line or tower inclination, vertical line motion under windy conditions or ice accumulation, etc

13.2.6 The voltage and currents induced on an overhead line due to a nearby positive lightning downward leader in the presence of a distribution transformer and surge arrester

Lightning is a natural source of electrical interference. An important aspect of this interference is the determination of electromagnetic coupling to overhead lines and the response of the lines illuminated by an incident electromagnetic wave. Most studies on lightning-induced over-voltages [5, 7, 26–33] have used only the return stroke electric field changes as a source, neglecting any field changes prior to it. Moreover, they usually deal only with negative polarity. However, recent measurements [7, 23, 32] show that the leader electric field changes at very close distances are of the same order as that of the return stroke. Thus, some lightning protection devices could be operated before the return stroke.

Using a coupling model based on the transmission line theory for the field–overhead line coupling calculation, the transient voltages induced on a line of finite length were determined, assuming different strike points in the vicinity of the line. In this section, a distribution transformer protected by surge arresters, each located at one of the line terminations, was taken into account. Their presence can be useful in improving the co-ordination of insulation and electromagnetic compatibility studies.

The analysis was applied to a basic case that is described hereafter. Consider a cloud–ground distance of $D = 5000$ m, assuming a uniform field at the ground level of about 15–20 kV m^{-1} before the preliminary discharge of the voltage difference between the cloud and ground can be estimated to a few tens of MV. A gap length D was subjected to a constant voltage of about $U_c = 60$ MV under specific atmospheric conditions: $P = 100$ kPa and $T = 21°C$ at ground level. As concerns the cloud diameter, a kilometre constitutes a good dimension order [1, 21, 36, 37]. A distribution line with values of length $L_x = 0.5$ km, height = 10 m, terminated on a distribution transformer (protected by a surge arrester) at the left termination point A of figure 13.2 is considered. Point B is terminated by a resistance of about 500 Ω (which corresponds to the characteristic impedance of the line when considered lossless).

In the literature (e.g. [38]), different equivalent circuits have been proposed to model the transformer in impulse regimes. For the sake of simplicity, the equivalent circuit proposed by Iorio *et al* [39] was used; the transformer is modelled by means of an equivalent capacitance C of 1 nF. Note that such a capacitance may also represent a short cable connected to the line [39]. The surge arrester (put in parallel with the transformer) is represented by a non-linear resistance; its characteristics are shown in figure 13.14 [39].

Here we evaluate the induced effects of the leader on a nearby overhead line. The field–line coupling simulations have been performed for a 0.5 km long, 10 m high

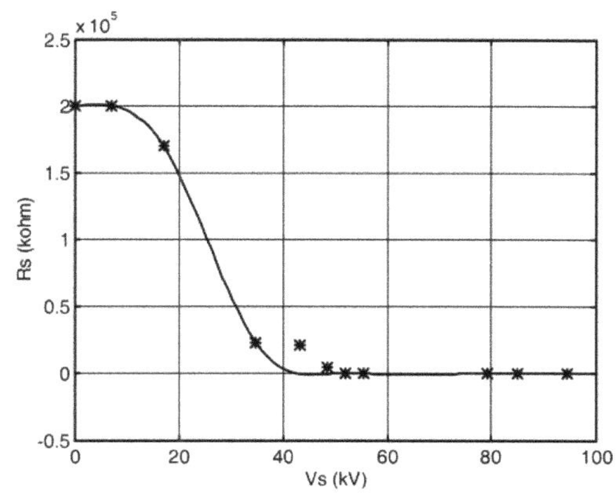

Figure 13.14. Assumed non-linear characteristics of the surge arrester, from [39].

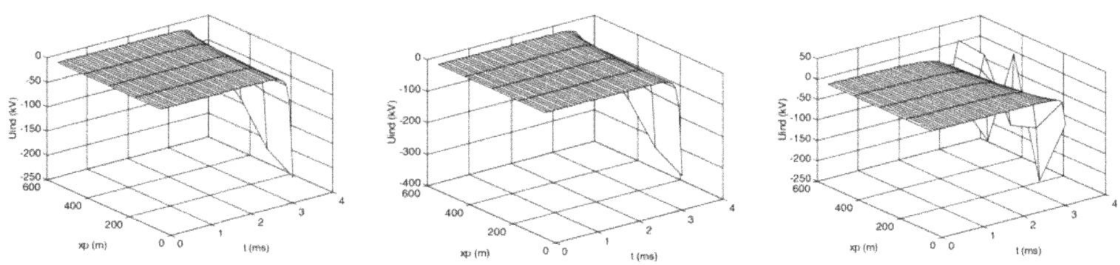

Figure 13.15. Induced voltage along the line for location 3: (a) line matched at both ends, (b) presence of transformer in A, and (c) presence of transformer protected by a surge arrester in A.

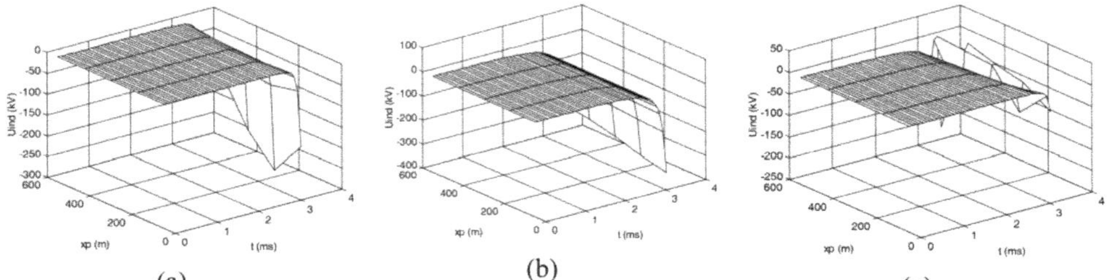

(a) (b) (c)

Figure 13.16. Induced voltage along the line for strike location 4: (a) line matched at both ends, (b) presence of transformer in A, and (c) presence of transformer protected by a surge arrester in A.

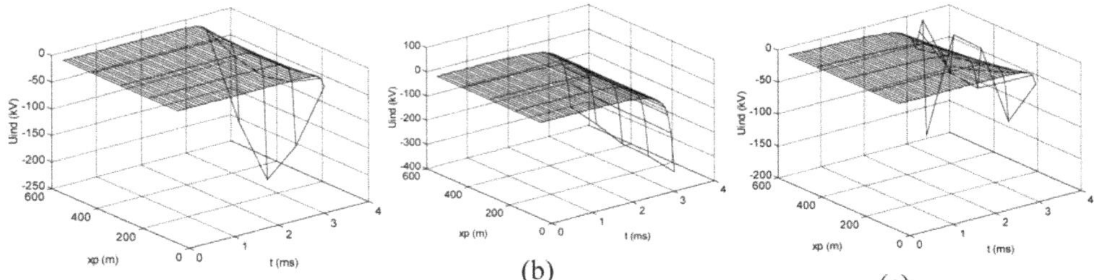

(a) (b) (c)

Figure 13.17. Induced voltage along the line for strike location 5: (a) line matched at both ends, (b) presence of transformer in A, and (c) presence of transformer protected by a surge arrester in A.

distributed line. The wire and the ground were assumed to be perfect conductors. Several strike locations, indicated in figure 13.4, have been considered. The induced voltages have been computed at five equidistant points along the line for only three locations (3, 4, and 5 of figure 13.4). The simulation results for three cases (matched line, presence of transformer, and presence of a transformer protected by a surge arrester) are shown in figures 13.15–13.17.

The surge arrester reacts as a reflected travelling wave propagating along the line to reduce the induced over-voltage. It may be noticed that the highest peak values occur in the case of an un-protected transformer.

References

[1] Uman M A 1987 *The Lightning Discharge* (International Geophysics Series vol 39) (New York: Academic)

[2] Ben Rhouma A 1996 Caractérisation et modélisation des champs électromagnétiques rayonnés par les décharges de foudre en vue de l'élaboration d'un simulateur expérimental *PhD Thesis* Ecole Centrale de Lyon, University of Lyon

[3] Zeddam A and Degauque P 1987 *Electromagnetics* **7** 541

[4] Degauque P and Zeddam A 1988 *IEEE Trans. EMC* **30** 77

[5] Diendorfer G 1990 *IEEE Trans. EMC* **32** 292

[6] Nucci C A, Rachidi F, Ianoz M V and Mazzetti C 1993 IEEE. Trans. EMC **35** 75

[7] Rachidi F, Rubinstein M, Nucci C A and Guerrieri S 1994 Influence of the leader electric field change on voltages induced by very close lightning on overhead lines *22nd Int. Conf. Lightning Protection* (*Hungary, Budapest 19–23 September*) paper R 3b-10

[8] Tesche F M 1995 Principles and applications of EM field coupling to transmission lines *Proc. Suppl. 11th Int. Zurich Technical Exhibition on EMC* (*Zurich, 7–9 March*) tutorial T2 p 21

[9] Rachidi F, Nucci C A, Ianoz M V and Mazzetti C 1996 *IEEE Trans. EMC* **38** 250

[10] Guerrieri S, Ianoz M, Rachidi F, Nucci C A and Mazzetti C 1996 Lightning induced over-voltages on an overhead line above lossy ground: a sensitivity analysis *Proc. 23rd Int. Conf. Lightning Protection* (*Firenze, Italy, 23–27 September*) **vol 1** p 328

[11] Ianoz M 1997 Lightning electromagnetic effects on lines and cables *12th Int. Zurich Symp. and Technical Exhibition on EMC* (*Zurich, Switzerland, 18–20 February*) tutorial T3, p 39

[12] Rubinstein M, Rachidi F, Uman M A, Thottapillil R, Rakov V A and Nucci C A 1995 *J. Geophys. Res.* **100** 8863

[13] Fofana I 1996 Modeling of positive discharge in long air gaps *PhD Thesis* Ecole Centrale de Lyon, University of Lyon

[14] Rubinstein M, Uman M A, Medelius P J and Thomson E M 1994 *IEEE Trans. EMC* **36** 134

[15] Master M J, Uman M A, Beasley W H and Darveniza M 1986 *IEEE Trans. EMC* **28** 158

[16] Cooray V 1994 *IEEE Trans. EMC* **36** 179

[17] Chang-Chou Hwang 1997 *IEEE Trans. Magn.* **33** 1520

[18] Michishita K, Ishii M and Imai Y 1996 Lightning-induced voltages on multi-conductor distribution line influenced by ground conductivity *23rd Int. Conf. on Lightning Protection* (*Firenze, Italy, 23–27 September*) vol 1 p 30

[19] Agrawal A K, Price H J and Gurbaxani S H 1980 *IEEE Trans. EMC* **22** 119

[20] Iorio R, Nucci C A, Porrino A and Rachidi F 1993 Lightning-induced over-voltages on distribution overhead lines in presence of distribution transformers or short cables *Int. CIGRE Symp. Power Syst. Electromag. Compatilibility* - Lausanne

[21] Fofana I and Beroual A 1998 *Electr. Power Syst. Res* **48** 105

[22] Aguet M and Ianoz M 1987 *Haute Tension* (Paris: Dunod)

[23] Thomsom E M, Uman M A and Beasly W H 1980 *J. Geophys. Res.* **90** 8136

[24] Rakov V A, Uman M A, Fernandez M I, Thottappillil R, Eybert-Berard A, Berlandis J P, Rachidi F, Rubinstein M, Guerrieri S and Nucci C A 1996 Observed electromagnetic environment close to the lightning channel *Proc. 23rd Int. Conf. on Lightning Protection* (*Firenze, Italy, 23–27 September*) p 30

[25] Georgiadis N, Rubinstein M, Uman M A, Medelius P J and Thomson E M 1992 *IEEE Trans. Electromagn. Compat.* **34** N451

[26] Yokohama S 1984 *IEEE Trans. Power Appar. Syst* **103** 100

[27] Cooray V and De La Rosa F 1986 *IEEE Trans. Antennas Propag* **34**

[28] Zeddam A and Degauque P 1990 Current and voltage induced on telecommunication cable by a lightning stroke *Lightning Electromagnetics* ed R L Gardner (Washington, DC: Hemisphere) p 377

[29] Rubinstein M, Tzeng A Y, Uman M A, Medelius P J and Thomson E M 1989 *IEEE Trans. EMC* **31** 376

[30] Rachidi F, Nucci C A, Ianoz M V and Mazzetti C 1996 *IEEE Trans. EMC* **38** 250

[31] Cooray V 1995 *IEEE Trans. Power Deliv* **10** 330

[32] Uman M A, Rakov V A, Rambo K J, Vaught T W, Fernandez M I, Cordier D J, Chandler R M, Bernstein R and Golden C 1997 *Trans. IEEE Japan* **117B** 446

[33] Ishii M, Michishita K, Hongo Y and Oguma S 1994 *IEEE Trans. Power Deliv* **9** 109
[34] Beger K 1967 J *Franklin Inst.* **283** 478
[35] EPRI 2014 *EPRI AC Transmission Line Reference Book: 200 kV and Above* (Palo Alto, CA: EPRI) *ID* 3002003747
[36] Fofana I, Beroual A, Rhouma B A and Auriol P 1998 *IEEE Power Gen. Trans. Syst* **145** 395
[37] Fofana I and Beroual A 1997 *J. Phys. D: Appl. Phys* **30** 1653
[38] Voivre V, Arthaud J P, Ahmad A and Burais N 1993 *IEEE Trans. PWDR* **8** 212
[39] Iorio R, Nuci C A, Porrino A and Rachidi F 1993 Lightning-induced over-voltages on distribution overhead lines in presence of distribution transformers or short cables *Symp. CIGRE, EMC in Power Networks* (*Lausanne, Switzerland, 18–20 October*) paper 300-09

IOP Publishing

Discharge in Long Air Gaps
Modelling and applications
A Beroual and I Fofana

Chapter 14

Negative lightning model—applications

14.1 Introduction

Negative lightning discharges have been the subject of very intensive experimental investigation. In contrast to positive lightning, research to derive models for negative lightning is still in its early stages. This is mainly due to the complexity of the mechanisms involved in the propagation processes. Similarly to positive lightning discharges, the prediction of the parameters characterizing a negative lightning discharge and the return stroke is of a great interest for the protection of people and structures. Recall that about 80%–90% of all lightning strikes in temperate regions are of negative polarity.

In chapter 8, a predictive dynamic model of negative lightning discharge was proposed to predict its main macroscopic parameters. In this chapter simulation results obtained from this mathematical model are reported along with some engineering applications. The applications performed with positive polarity (chapters 11–13) can be performed under negative polarity, based on the same concepts. In this chapter, only a few applications are reported. In addition to lightning parameters (discharge current, propagation velocity, etc), radiated electric and magnetic field changes are presented.

14.2 The prediction of negative lightning discharge parameters

For the validation of the model, three cloud–ground distances were considered, namely 2000, 5000, and 7500 m subjected to 35, 45, and 50 MV, respectively, under the specific atmospheric conditions of $P = 100$ kPa and $T = 20$ °C at ground level. These values are in the range of those used in the literature [1–4]. The base of the cumulonimbus cloud was considered as cylindrical disk 1 km in diameter. The initial radius r_0 of the negative leader is taken to be equal to 0.75 cm [1].

Figures 14.1–14.3 show some simulated results of the downward negative lightning leader characteristics. Example of trajectories are given in figures 14.1(a), 14.2(a), and 14.3(a). From those results, it can be observed that the simulated current

doi:10.1088/978-0-7503-1236-3ch14

(figures 14.1(b), 14.2(b), and 14.3(b)), the charge (figures 14.1(d), 14.2(d), and 14.3(d)) and energy injected into the gap (figures 14.1(c), 14.2(c), and 14.3(c)) depict a real stepping. This is a well-known phenomenon concerning the propagation of negative lightning. The current values are in the same range as those observed experimentally [1–3]. Indeed, the simulated pulse currents of steps can reach a few kA. According to experimental results reported in the literature, the currents present stepwise pulses of at least 1 kA [3]. The current pulses correspond to re-illuminations. The number of these re-illuminations depends on the voltage, the gap length, and hence on the energy injected in the gap. It increases non-linearly with the cloud–ground gap. Moreover, the repetition time of stem occurrence lies between 0.1 and 1 μs, while the time duration between two re-illuminations lies between 0.125 and 0.5 ms (figures 14.1(b), 14.2(b), and 14.3(b)). These results are of the same order as those estimated experimentally [1–3]. The electrical charge injected during the propagation can reach several coulombs. The simulated stepped leader duration seems to be higher—by about one order of magnitude—than the average one obtained experimentally. The characteristics of the

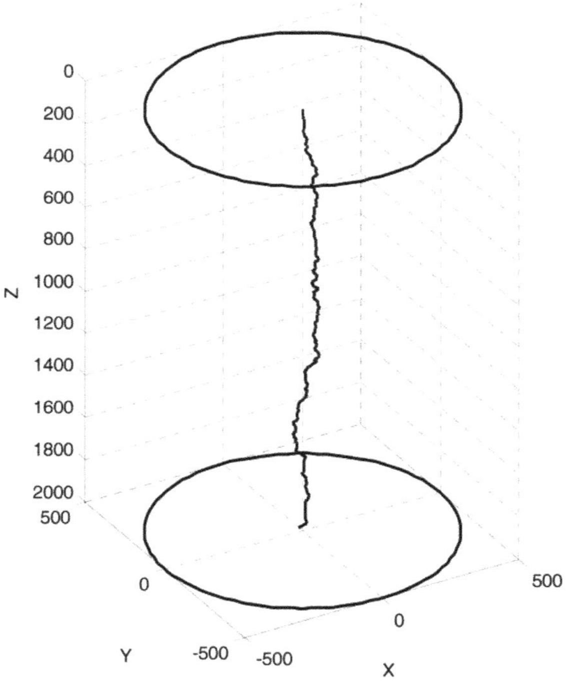

(a)

Figure 14.1. Simulation of cloud–ground distance D = 2000 m under a constant voltage U_c = 35 MV, and atmospheric conditions of P = 100 kPa and T = 20 °C. The initial radius of the equivalent electrode r_i = 6 m [1]. (a) Simulated main leader's trajectory in space. (b) Simulated main leader's current. (c) Simulated energy injected into the gap. (d) Applied voltage between cloud–ground. (e) Simulated charge injected into the air gap. (f) Main leader thermal radius evolution. (g) Main leader instantaneous velocities. (h) Gradient in the main leader. (i) The simulated discharge's return strike current.

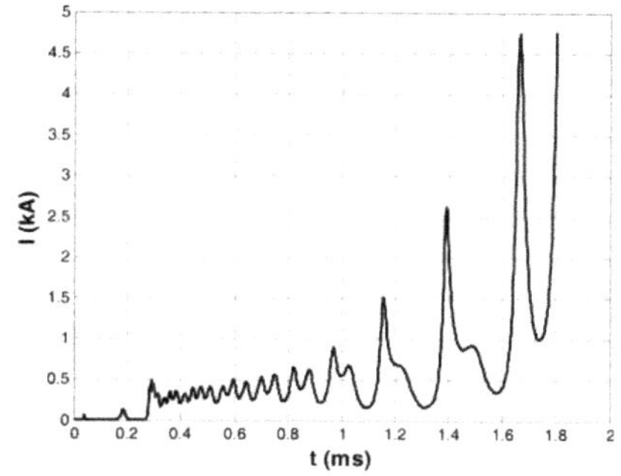

(b)

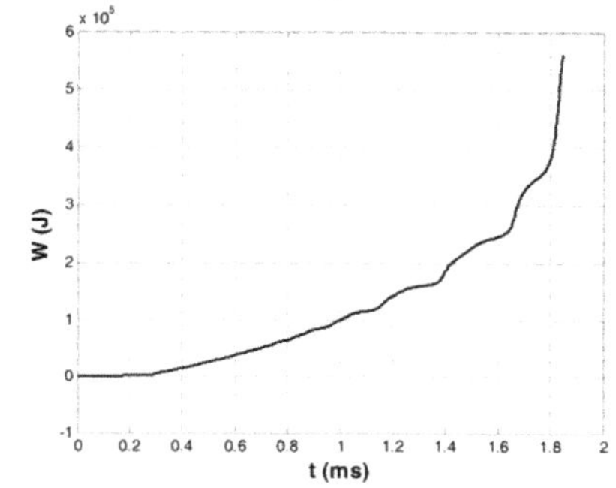

(c)

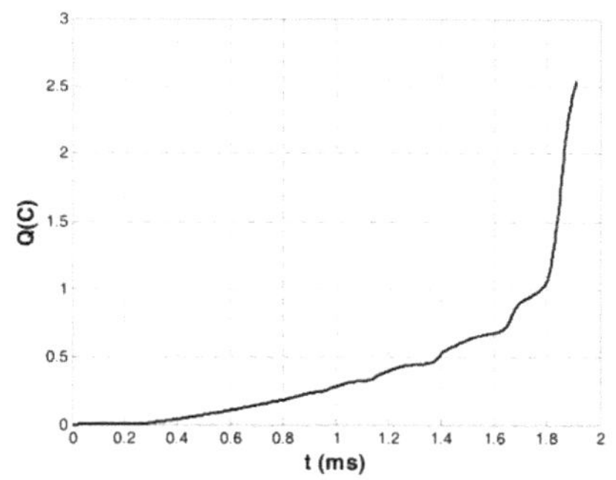

(d)

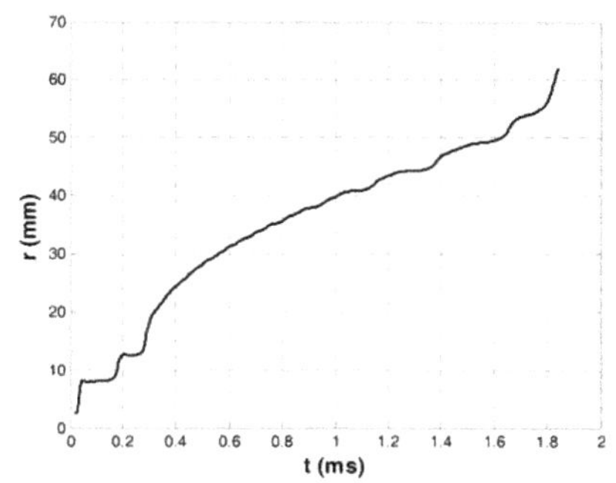

(e)

Figure 14.1. (Continued.)

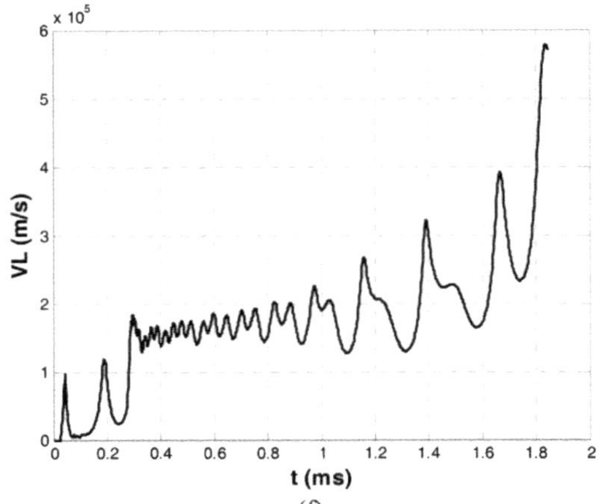

(f)

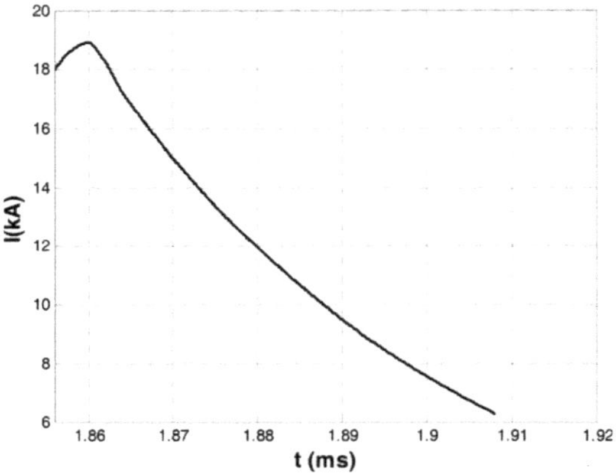

(g)

Figure 14.1. (Continued.)

thermal radius are reported in figures 14.1(e), 14.2(e), and 14.3(e). It grows by a few centimetres as observed for natural lightning [1, 3]. It varies with the injected charge; the higher the injected charge, the larger the radius is. The simulated leader instantaneous velocities are a few 10^5 m s^{-1} (figures 14.1(f), 14.2(f), and 14.3(f)) and the simulated average value is about 3×10^5 m s^{-1}. Such a value is in good agreement with that obtained experimentally from streak camera recordings, at the bottom of the stepped leader's channel (i.e. near the ground) during natural lightning discharge [1, 2, 5–8]. Rubinstein *et al* [5] obtained 4.0×10^5 m s^{-1} near the cloud base and 5.2×10^6 m s^{-1} near the ground with an average value of about 2×10^6 m s^{-1}. Figures 14.1(g), 14.2(g), and 14.3(g) depict the predicted return stroke currents. The peak value of the return stroke can reach 18.5, 20.5, and 30.5 kA for 2000, 5000, and 7500 m, respectively. Such values are in the range of those reported in the literature [1, 2]. For instance, the mean

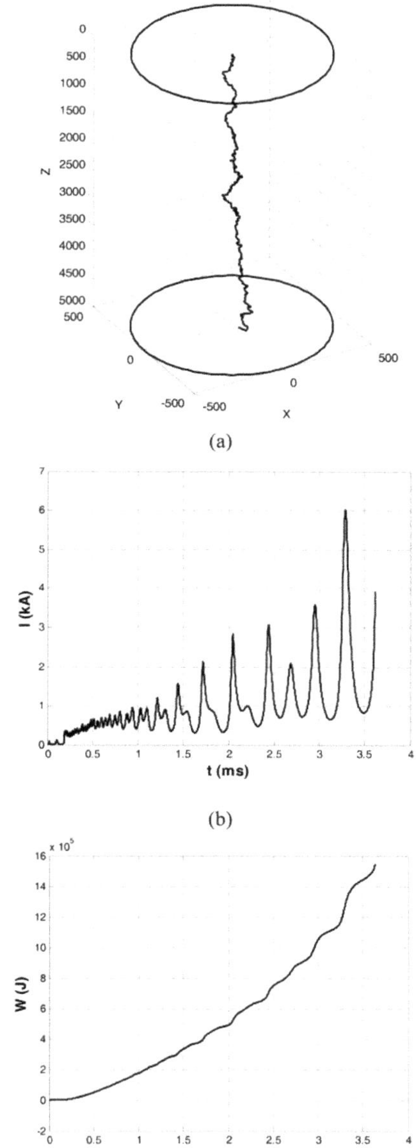

Figure 14.2. Simulation of cloud–ground distance $D = 5000$ m under a constant voltage $U_c = 45$ MV, and atmospheric conditions of $P = 100$ kPa, and $T = 20$ °C. The initial radius of the equivalent electrode $r_i = 6$ m [1]. (a) Simulated main leader's trajectory in space. (b) Simulated main leader's current. (c) Simulated energy injected into the gap. (d) Applied voltage between cloud–ground. (e) Simulated charge injected into the air gap. (f) Main leader thermal radius evolution. (g) Main leader instantaneous velocities. (h) Gradient in the main leader. (i) The simulated discharge's return strike current. © 2010 IEEE. Reprinted with permission from [9].

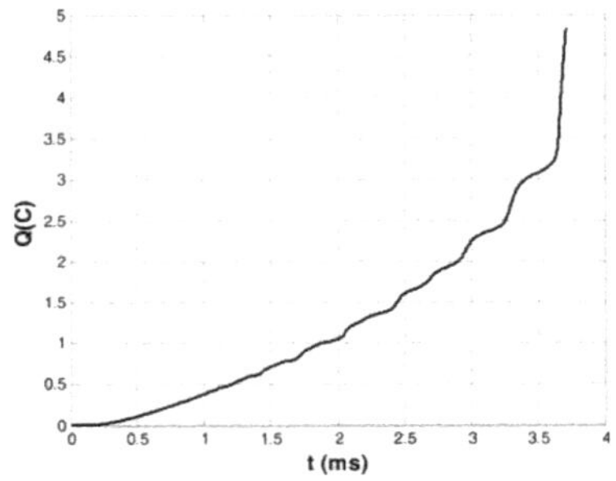

(d)

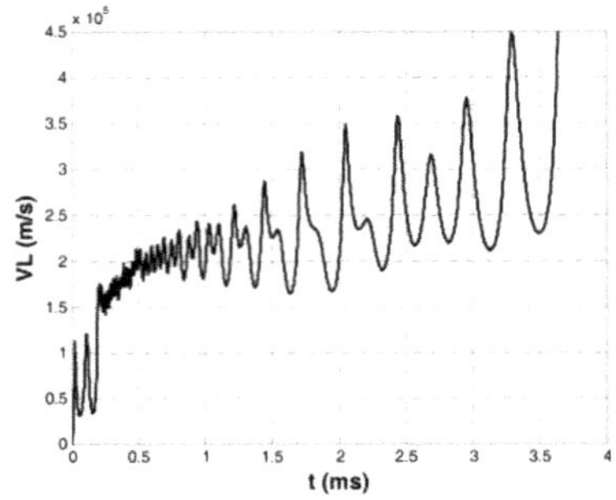

(e)

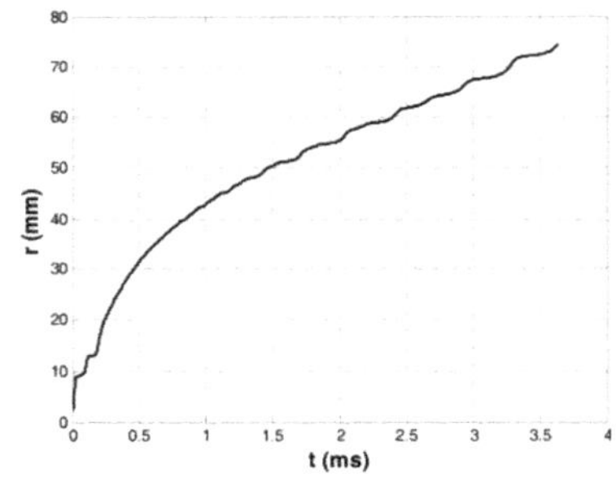

(f)

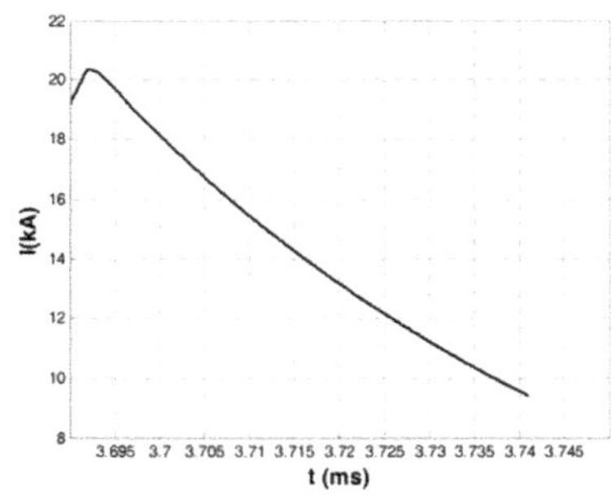

(g)

Figure 14.2. (Continued.)

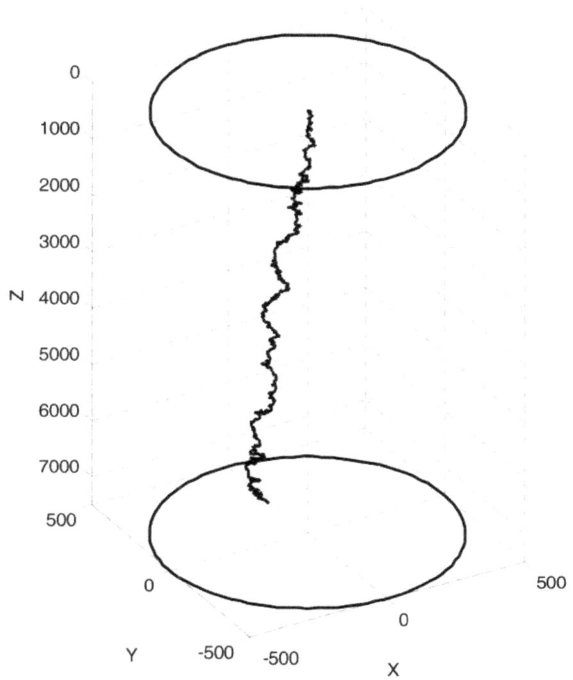

(a)

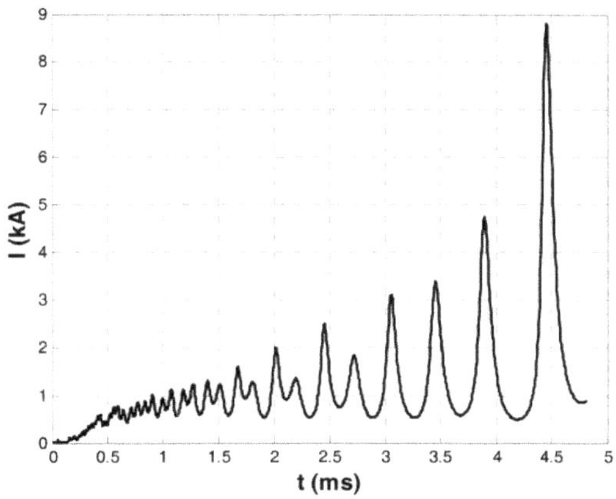

(b)

Figure 14.3. Simulation of cloud–ground distance $D = 7500$ m under a constant voltage $U_c = 50$ MV, and atmospheric conditions of $P = 100$ kPa and $T = 20$ °C. The initial radius of the equivalent electrode $r_i = 6$ m [1]. (a) Simulated main leader's trajectory in space. (b) Simulated main leader's current. (c) Simulated energy injected into the gap. (d) Applied voltage between cloud–ground. (e) Simulated charge injected into the air gap. (f) Main leader thermal radius evolution. (g) Main leader instantaneous velocities. (h) Gradient in the main leader. (i) The simulated discharge's return strike current. © 2010 IEEE. Reprinted with permission from [9].

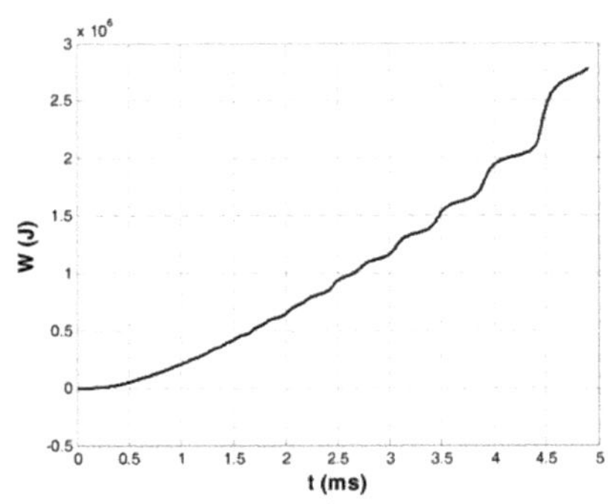

(c)

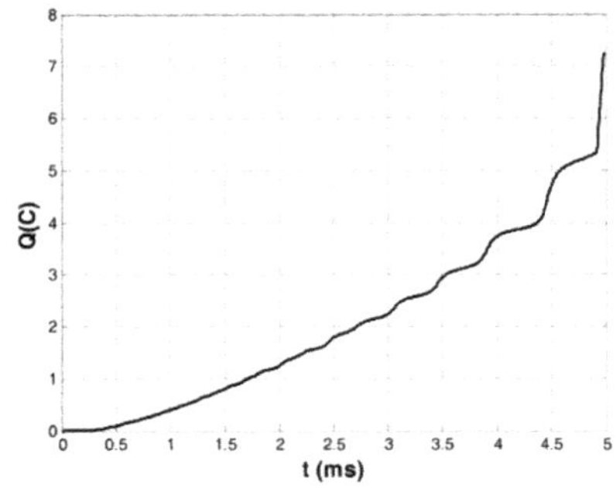

(d)

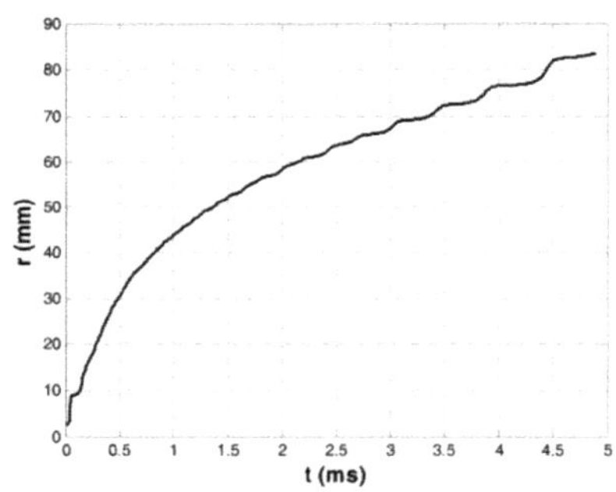

(e)

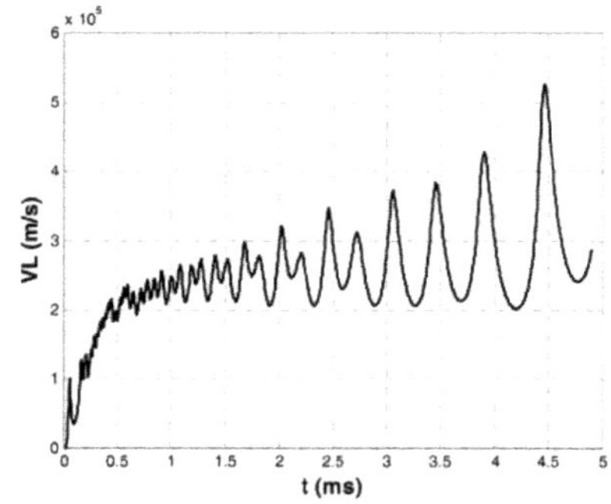

(f)

Figure 14.3. (Continued.)

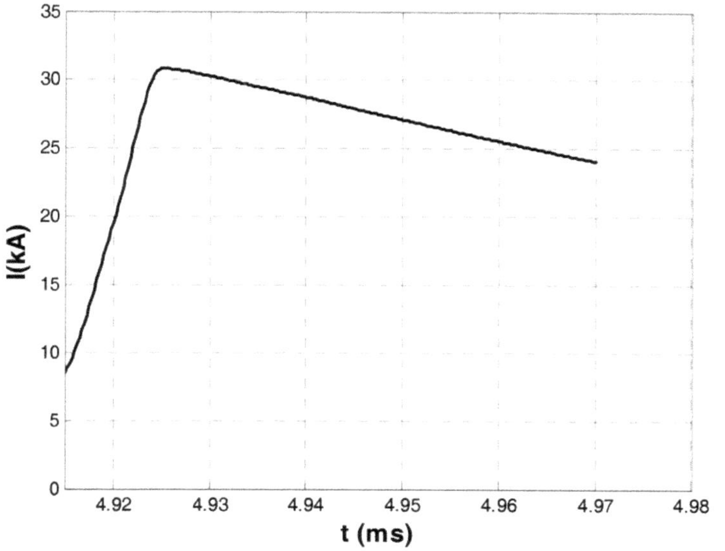

(g)

Figure 14.3. (Continued.)

Table 14.1. Simulated results for the three different cloud–ground distances.

Gap$D$ (m)	Crest voltage U_{cr} (MV)	Maximum velocity v_{max} (m s^{-1})	Leader average current I_m (kA)	Maximum current of return stroke I_{arc} (kA)
2000	35	4.0×10^5	1	18.5
5000	45	4.5×10^5	1.5	20.5
7500	50	5.2×10^5	2	30.5

first-stroke peak current measured by K Berger [10] for negative discharges on Mount San Salvatore (Switzerland) is 30 kA. Table 14.1 summarizes the main parameters obtained for the three cloud–ground distances.

For a given voltage, the relationship giving the velocity of negative and positive discharges [11–13] shows that the current (i_L) and velocity (v_L) of the leader are related as $v_L \sim i_L^{1/3}$. A fairly close relationship ($v_L \sim i_L^{0.30}$) was deduced by a number of researchers from their experimental observations [8, 14, 16].

The proposed model may be considered as a good prediction tool for the characteristics of negative lightning discharges. The simulated temporal evolution of the negative lightning discharge depicts the step-like propagation as well as the different sequences of the propagation as reported in the literature. Moreover, the simulated values of the characteristic parameters are generally in the same range as the experimental ones estimated during natural lightning and reported in the literature. The model also emphasizes the randomness aspect of the negative leader trajectory (figures 14.1(a), 14.2(a), and 14.3(a)). Indeed, for the same input parameters, one can have different simulated trajectories. Since the results vary randomly from one simulation to the next, one cannot expect precise agreement between a single run of the simulation and a given experimental discharge.

14.3 Electric and magnetic fields associated with the leader

The electric and magnetic fields radiated by the negative downward leader were computed for two cloud–ground distances, namely, 5000 and 7500 m. Using the set of equations presented in section 11.4, the electric and magnetic field changes radiated by the discharge seen at an observation distance *Dh* were predicted. Figures 14.4–14.7 represent the electric and magnetic fields simulated for the conditions listed above. The amplitudes and the evolution of these fields are in good agreement with those observed experimentally during the artificially triggered lightning campaign at Camp Blanding, Florida, between 1993 and 1995 [15].

The overall computed space–time characteristics of the negative lightning discharge show satisfactory agreement with experimental measurements and observations. They prominently display the scale effects between the simulated results obtained for negative lightning discharges and long negative discharges in the laboratory. The proposed simulator may, therefore, constitute a good tool in determining the characteristics of negative lightning discharge. This work

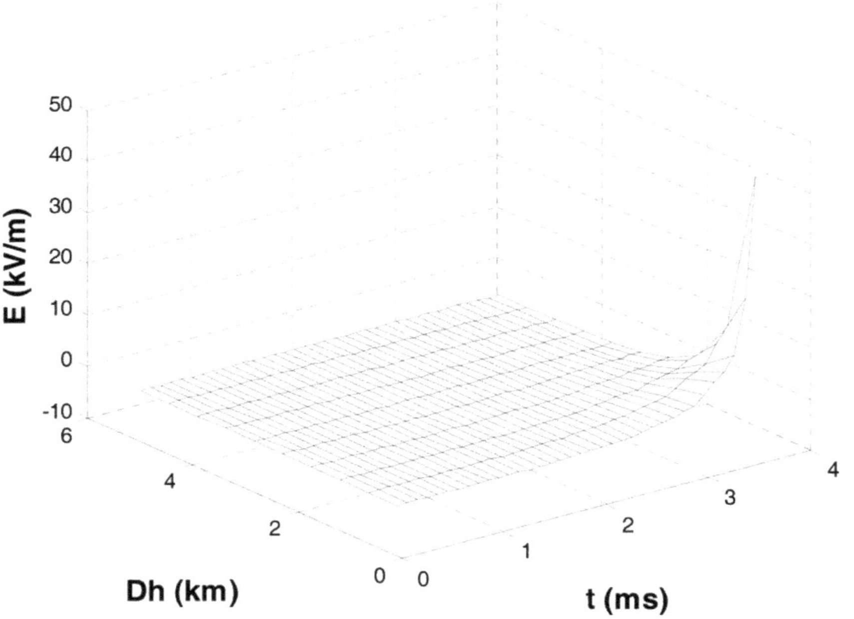

Figure 14.4. The leader electric field at different positions from the discharge before the final jump. Cloud–ground distance = 5000 m and voltage = 45 MV.

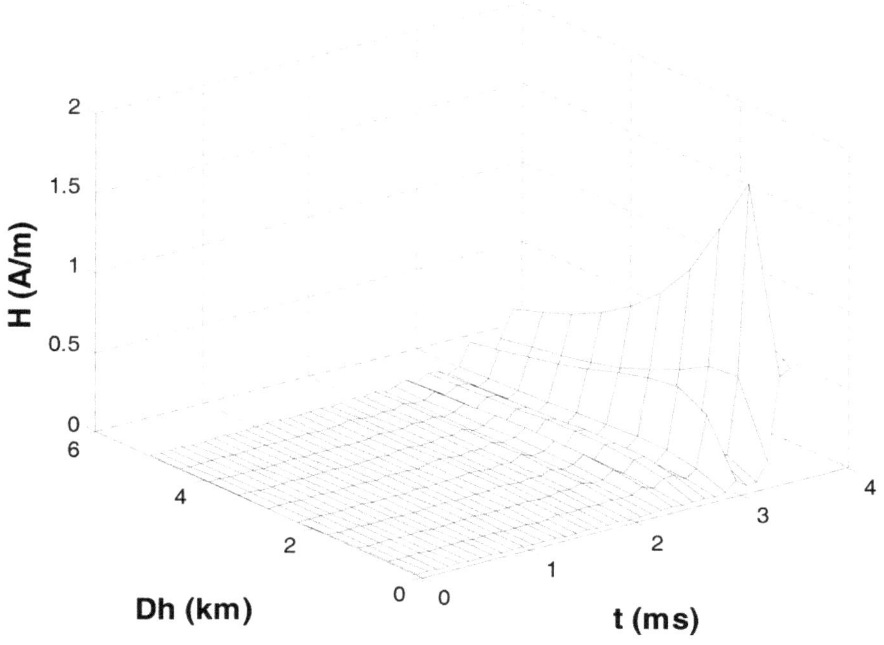

Figure 14.5. The magnetic field radiated by the negative electrode leader at different positions from the discharge before the final jump. Cloud–ground distance = 5000 m and voltage = 45 MV.

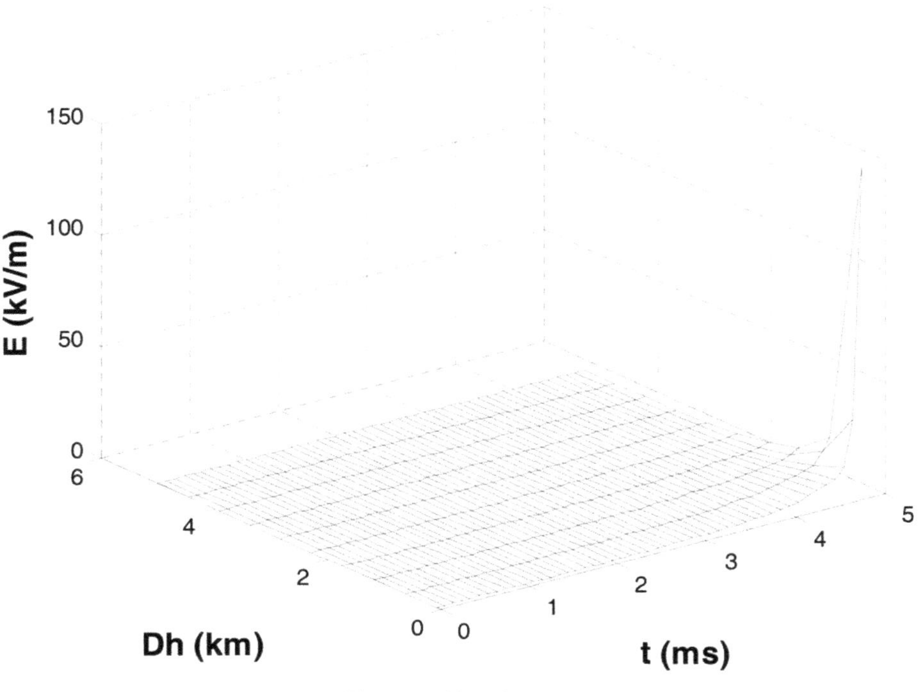

Figure 14.6. The leader electric field at different positions from the discharge before the final jump. Cloud–ground distance = 7500 m and voltage = 50 MV.

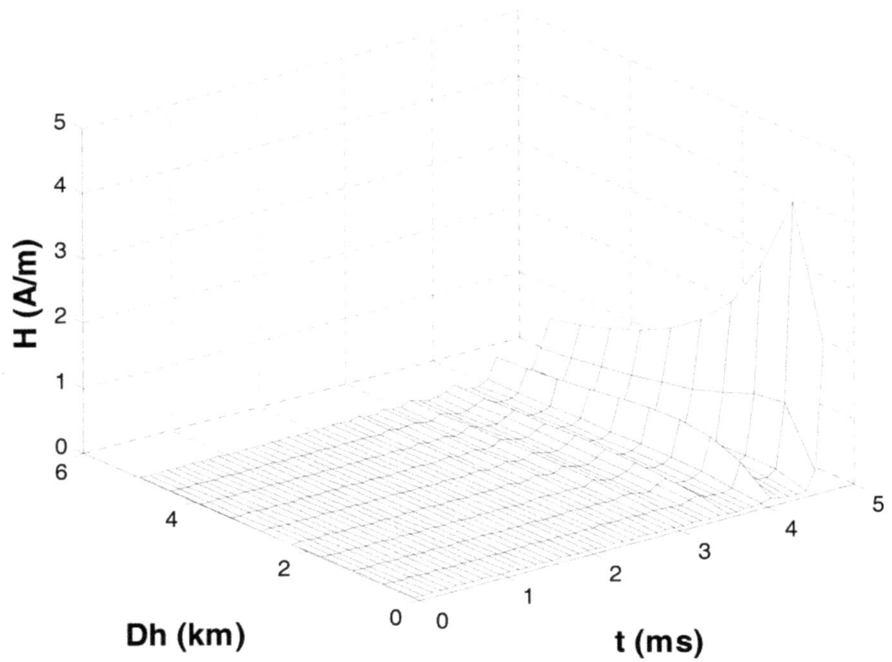

Figure 14.7. The magnetic field radiated by the negative electrode leader at different horizontal positions from the discharge before the final jump. Cloud–ground distance = 7500 m and voltage = 50 MV.

emphasizes once more the utility of equivalent electrical circuits in modelling physical phenomena. The presented results have shown that the actual understanding of the physical mechanism of electrical discharge is sufficient for the development of self-consistent detailed models. Work is still in progress to improve the values of the physical parameters and to check the limits of accuracy over a wide range of experimental results. The authors are currently engaged in additional work to study the influence of particular assumptions made on the lightning discharge parameters.

References

[1] Schonland B F J 1956 *The Lightning Discharge (Handbuch der Physic vol 22)* p 576

[2] Uman M A 1987 *The Lightning Discharge (International Geophysics Series vol 39)* (New York: Academic)

[3] Le Roy G, Gary C, Hutzler B, Lalot J and Dubanton C 1984 *Les Propriétés Diélectriques de l'Air et les Très Hautes Tensions (Collection de la Direction des Études et Recherches d'Électricité de France vol 51)* (Paris: Eyrolles)

[4] Bacchiega G L, Gazzani A, Bernardi M, Gallimberti I and Bondiou-Clergerie A 1994 Theoretical modelling of the laboratory negative stepped leader 16th *Int. Aerospace and Ground Conf. on Lightning and Static Electricity (Mannheim, Germany, May* 24–27)

[5] Rubinstein M, Rachidi F, Uman M A, Thottapillil R, Rakov V A and Nucci C A 1995 *J. Geophys. Res.* **100** 8863

[6] Orville R E and Idone V P 1982 *J. Geophys. Res.* **87** 11177–92
[7] Thomson E M, Uman M A and Beasley W H 1985 *J. Geophys. Res.* **90** 8136
[8] Bazelyan E M and Raizer Y P 2000 *Lightning Physics and Lightning Protection* (Bristol: Institute of Physics Publishing)
[9] Beroual A, Rakotonandrasana J-H and Fofana I 2010 *IEEE Trans. Dielect. Elect. Ins.* **17** 1562–8
[10] Beger K 1967 *J. Franklin Inst.* **283** 478
[11] Beroual A 2004 *REE Rev. Electr. Electron.* **8** 36
[12] Beroual A 1993 *J. Appl. Phys.* **73** 4528
[13] Fofana I and Beroual A 1996 *J. Phys. D: Appl. Physics* **29** 691
[14] Rizk F A M 2008 *IEEE Trans. Power Deliv.* **23** 296
[15] Uman M A, Rakov V A, Rambo K J, Vaught T W, Fernandez M I, Cordier D J, Chandler R M, Bernstein R and Golden C 1997 *Trans. IEE Japan* **117B** 446
[16] Hutzler B and Hutzler D 1982 *Bull. Dir. Etud. Rech.* **4** 11

www.ingramcontent.com/pod-product-compliance
Ingram Content Group UK Ltd.
Pitfield, Milton Keynes, MK11 3LW, UK
UKHW051857100926
8556IPUK00005B/660